Sustainability in Plant and Crop Protection

Volume 15

Series Editor

Aurelio Ciancio, Sezione di Bari, Consiglio Nazionale delle Ricerche Istituto per la Protezione delle Piante, Bari, Italy

The series describes new perspectives recently developed in pest and disease management, based on innovative tools emerging from basic and applied research. The volumes will aim at interested readers involved in plant protection and crop management, for whom soil biodiversity, crop sustainability and, in general, organic approaches are fundamental. Different cropping systems will be treated by researchers involved in cutting edge studies worldwide. A number of basic issues including sustainability, life-cycle assessment, evolution, plant nutrition and organic cropping technologies will provide a common framework, within which different components of the crop production cycle will be focused on. These will range from roots and endophytes to pest and disease control, through the management of soil microbiome and fertility. These issues will be examined at the field and crop levels, including the effects of invasive species and climate changes on agroecosystems. Recent advancements in massive sequencing will represent the basis of dedicated volumes, dealing with transcriptomics and related approaches. They will illustrate the potentials and benefits of extensive DNA and RNA data analyses and studies, for practical purposes in crop protection, disease management and food production.

Contributions on any of the above cited topics are welcome. Potential Editors proposing a new volume are requested to contact the Series Responsible Editor and provide a short CV (400 words max) listing a selection of their most significant publications. In order to broaden the base of contributors and avoid redundancies, only one volume per Editor is allowed. Exceptionally, in case of many contributed chapters, a two-issues volume can eventually be considered.

More information about this series at http://www.springer.com/series/13031

Juan F. Barrera

Beyond IPM: Introduction to the Theory of Holistic Pest Management

Springer

Juan F. Barrera
El Colegio de la Frontera Sur
Tapachula, Chiapas, Mexico

ISSN 2567-9805 ISSN 2567-9821 (electronic)
Sustainability in Plant and Crop Protection
ISBN 978-3-030-43369-7 ISBN 978-3-030-43370-3 (eBook)
https://doi.org/10.1007/978-3-030-43370-3

This Springer imprint is published by the registered company Springer Nature Switzerland AG.
The registered company address is: Gewerbestrasse 11, 6330 Cham, Switzerland

To my wife, Mara
To my daughter, Mara Alejandra, and my sons, Jupiter Raúl, Víctor Hugo, and Juan Gerardo
To my parents, Júpiter Barrera Flores (RIP) and Norma Gaytán Aguirre (RIP)

Foreword by Aurelio Ciancio

In the present volume, the author proposes and describes a new paradigm in pest management. This is the Holistic Pest Management (HPM) approach of which the author is a firm and enthusiastic supporter and fine scientific investigator. HPM basic concept relies on placing the farmer's well-being at the center of any pest management action. This is deployed through the identification and application of several parameters and variables for risk assessment and socioeconomic analyses. The HPM theory by itself is fascinating, not only because it affords the problem of linking pest management to environmental protection but mainly because of its focus. Attention is in fact given to the farmer and social aspects of crop production rather than to the crop productivity alone, considering income and yields as a way (and not as a goal) to receive a benefit from a more socially equal agriculture.

A number of questions, however, arise from the adoption of this perspective. In particular, can the holistic approach be applied to any socio-agrosystem? Moreover, may a new determinism in socio-ecology provide a suitable new philosophy capable to sustain, through its practical application, a crowded planet? Can we leave apart many aspects of complexity not yet considered in HPM, including the effect of chaotic and fast changes of food webs (including climate effects), and at which scales can HPM be applied?

The author is aware of these aspects and correctly starts his exposition by providing data on global agricultural ecosystems, including crop production, emerging issues in plant protection, and socioeconomic scenarios that may determine the success or failure of the management options chosen. This is followed by a historical review of conventional IPM through an epistemological and updated approach, which is useful both for students and researchers. In the following chapters, more details on HPM are progressively added, moving the reader from theory to practice, by means of a detailed methodology description, with the help of many graphic illustrations, and description of the author's own experience.

As a direct consequence of the HPM paradigmatic shift, new fields of study also emerge. The integration of factors external to the farm but directly involved in the decision-making process concerning the pest requires a perspective based on a rural anthropology and sociology. New studies will be needed on the impact of social organizations including state structures, (micro-)credit providers, multinationals, and markets, as well as on the role of predatory or marginal agriculture. All these aspects lead us to consider again the complexity of the world agriculture and the uncertainties linked to such a fascinating approach.

The new perspectives emerging by the paradigmatic shift herein proposed need to be consolidated and tested in the future. When passing from theory to reality, more practical examples will be needed to test HPM and to get a better comprehension of its implications and value. One of the recurrent questions posed, "why a pest is a pest," will surely deserve more analyses. It calls for studies integrating the evolutionary biology of species with the effects of anthropic changes such as deforestation, intensive agriculture, and monoculture, finally considering globalization and trade.

The volume is rich in citations, and the text is well written, correlated and supported by a comprehensive description of the work of many investigators, analyzed in detail. The author has the merit of offering the reader a new way of thinking and seeing at an old problem. He has to be acknowledged for dedicating such an effort to the divulgation of this new perspective and methodological approach, which will undoubtedly raise interest in the future, by promoting further research and applications.

IPSP, CNR
Bari, Italy

Aurelio Ciancio

Foreword: Holistic Pest Management by Pablo Liedo

Humankind has suffered the effects of pests since ancient times. Since then, control methods were sought and developed to suppress pest populations and reduce their damage. From the mid-1940s of the last century, through advances in synthetic chemistry, highly effective and cheap insecticides were developed. Their impact was such that pests were considered history of the past. The negative effects on the environment and beneficial organisms, the accumulation of toxic waste, and the emergence of resistance and secondary pests were phenomena that were not given importance. The consequence of dependence on pesticides was, in the best scenario, the search for more effective chemicals with a lower impact on the environment. This represented a significant increase in costs. At the other extreme, the abuse of insecticides resulted, in some cases, in an irreparable deterioration of nature and even in the loss of human lives due to poisoning.

Then, in the 1960s, it was concluded that it was necessary to use all available methods to prevent the damage of pests, leaving the use of insecticides as a last resort. Integrated Pest Management (IPM) became not only the desirable strategy but even a school of thought, the paradigm.

The concept of IPM and its philosophy has been widely accepted to such an extent that today there are IPM associations and many universities offer specialties or postgraduate degrees in this field. Many countries have identified it as the basis of their human, animal, and plant public health policies. Organizations such as FAO and WHO have adopted it as the strategy to address pests and disease vectors. This wide acceptance is due to the fact that the concept implies the optimization of resources, minimizing externalities or unwanted side effects. Put simply, IPM means "doing the right thing." Who could oppose this philosophy?

However, the question is not whether the concept is accepted or not but rather if it is applied or is only theory and good intentions.

There is evidence demonstrating the effectiveness of IPM, suppressing pests, reducing their damage, and minimizing the application of pesticides. However, there is also ample evidence that this strategy is not applied or adoption has been slow or that IPM are short-lived. What are the reasons for this?

It is from this reflection and field experience, when developing the technology for the integrated management of the coffee berry borer, that the idea of Holistic Pest Management (HPM) was brewing in Juan Barrera. The need for a paradigm shift puts human being at the center of a participatory decision-making system for pest management. As the name implies, its main characteristic is the holistic approach.

This book represents the culmination of an intellectual effort of 15 years, accompanied by a constant and close experience, experimentation, and discussion with rural producers, students, and fellow researchers. It starts with a review of the IPM concept, its background, and limitations. It raises and justifies the need for the holistic approach and its theoretical support in the science of complexity. In this volume, Barrera describes the methods for application and implementation of HPM, which have been tested and evaluated. Finally, he analyzes the perspectives and challenges of the new approach. I trust it will be a work of transcendence, not only in the field of pest management but also in the fields of agroecology, science, and education.

For me, it is an honor to foreword this book by my colleague and friend Juan Barrera, who has been trained at the IPM school under one of its founders in Mexico, Dr. Dieter Enkerlin. Over three decades, I have had the privilege of interacting with him in different projects and facets. I have always admired his high level of commitment and dedication, which has led him to fulfill his goals and objectives. This book is a clear example of that.

Tapachula, Chiapas, Mexico Pablo Liedo
December 2019

Preface

About 15 years ago, we asked ourselves why many Mexican coffee growers, particularly smallholders, did not use—or did so to a minimum—the integrated management of coffee berry borer (*Hypothenemus hampei*), the most important coffee crop pest worldwide. At that time, we deduced that this was due in large part to the fact that the control programs and campaigns were aimed at reducing the level of infestation but paid very little attention to reducing producer vulnerability and increasing their response capacity toward this pest.

When deepening the analysis of this situation, we found that the problem did not seem to be limited to Mexico, not even to coffee crop, that is, it was a much more widespread problem than previously thought. Therefore, we concluded that Integrated Pest Management (IPM) followed a typically reductionist strategy by focusing on the pest and detaching itself from the problems associated with other components of the crop production system, including the producers themselves.

When reviewing the literature on the subject, we found more criticism in other parts of the world, which made us suspect that something was lacking in IPM. The concordance of our ideas with those of other researchers reinforced our efforts to find a holistic approach to pest management decisions. As a starting point, we took two central ideas: first, the actions should always put the farmer at the center of the system, and, second, management should consider both pests and other system components. We have called this approach "Holistic Pest Management" (HPM). In order to publicize these concepts and the development of their advances, in this book, I present an introduction to the theory of HPM, a new paradigm that goes beyond IPM. It is not a matter of renaming or redefining the IPM but of replacing it.

After the introductory chapter (Chap. 1), where the foundations that justify proposing an approach to "go beyond IPM" are laid, the nature of IPM is first reflected upon (Chap. 2). As we enter IPM analysis, it will be seen that this is a decision-making system that promotes actions that lead to pest management under a reductionist approach. Hence, IPM reductionism appears as the main cause of its anomalies.

Subsequently, we will see that for a realistic management of pests, a holistic approach to them is essential (Chap. 3). At this point, we will venture into the philosophy of holism and the concepts related to complex systems management; we will exhaust the issue by analyzing some of the main alternatives to IPM that have emerged in the last 30 years.

In the succeeding chapters, the HPM is addressed. First, it begins with the concept, principles, and theory that underpin it (Chap. 4); next, the description of methods continues, emphasizing the procedures for calculating and interpreting the Holistic Risk Index (HRI)—the cornerstone of HPM—and how it relates to resilience (Chap. 5). Then, step-by-step and with some examples, the methodology to implement HPM is described (Chap. 6).

A final section (Chap. 7) compares IPM and HPM vis-à-vis to clear up their differences, including cases of HRI application in fields other than pest management, and finally the challenges of HPM.

I have no doubt that this book will be useful for both theorists and pest management practitioners. The book will even be of interest to those who study and manage complex systems from other disciplines.

Tapachula, Chiapas, Mexico — Juan F. Barrera
December 2019

Acknowledgments

My efforts to think and act holistically have been stimulated by more than 30 years of contact with coffee producers in Chiapas, Mexico. In particular, my thanks to Productores de Café La Central (PROCACEN) for their friendship and teachings. I appreciate the trust and collaboration of producers of the Centro de Agroecología San Francisco de Asís (CASFA), the Grupo de Asesores de Producción Orgánica y Sustentable S.C. (GRAPOS), and the Federación Indígena Ecológica de Chiapas (FIECH). Also, my recognition goes to the owners of coffee farms: Hamburgo, La Chiripa, Argovia, Linda Vista, Rancho La Esperanza, and Alianza.

Over the years, many ideas expressed herein about HPM were nurtured with the opinions and suggestions of Ramón Jarquín, Jürgen Pohlan, William Gamboa (RIP), Manuel Parra, Balente Herrera, Javier Valle, Jaime Gómez, Christiane Junghans, and Julio S. Bernal.

A special thanks goes to Balente Herrera, Jürgen Pohlan, Ramón Jarquín, Jaime Gómez, Christiane Junghans, Lorena Soto, Eduardo Bello, Erin Estrada, Vincenzo Bertolini, Miguel Ángel Castillo, Ernesto Sánchez, Graciela Huerta, and Isidra Ocampo, members of the Grupo de Investigación de ECOSUR en Zonas Cafetaleras (GIEZCA), with whom I have shared research projects, links with producers, and trainings of human resources since 2002, which have been an inspiration for the development of complex thinking.

With Edwin Castellanos, Ana Lucía Solano, Helda Morales, Hallie Eakin, Catharine Tucker, and Rafael Díaz, members of the "Cambios Globales y Café" project, I found the ideal interdisciplinary niche to implement some methods with a holistic approach.

In the company of Pablo Liedo and several generations of students of El Colegio de la Frontera Sur (ECOSUR) Doctoral Seminar on Pest Management, from 1998 to date, we review, comment, and discuss books and articles on various topics of Integrated Pest Management (IPM); thanks to this, the need to propose a paradigm shift germinated in me.

I want to thank Joel Herrera, Enrique López Pascacio, and Javier de la Rosa for their friendship and for their invaluable technical support in the field. I also appreciate the support received, at various times during the development of HPM, from

Javier Valle (statistical data analysis), Higinio López (analysis in Geographic Information Systems), Enoc Cabrera (Communication and Information Technologies), Raúl Cuevas (preparation of organic fertilizers), and Manuel Figueroa (linking with farmers).

Among those of my students who have dedicated their research thesis on pest management with producers, source of information for the HPM, I have to thank Ramón Jarquín, Héctor Segura, Lissette Rodríguez, Patricia Velazco, Julia Angélica Zavala, Walter Pereira, Carolina Reyes, and Luis Alberto Sumuano. I also thank my students Alejandra Serrano and Ariana Román for their support in data capture and ordering.

Much of the conceptual and methodological work on HPM arose from the following projects that were carried out between 2002 and 2018: "Bioecología y manejo de plagas del café en el Soconusco y Sierra de Chiapas" (Fundación Produce Chiapas, Mexico); "Promoting Organic Coffee Production in Chiapas Through Pest Management, Agronomic, and Economic Research" (Texas A&M University and Consejo Nacional de Ciencia y Tecnología, Mexico); "Plan Estatal de Manejo Agroecológico del Café en Chiapas" (Comisión para el Desarrollo y Fomento del Café en Chiapas, Mexico); "Effective Adaptation Strategies and Risk Reduction Towards Economic and Climatic Shocks: Lessons from the Coffee Crisis in Mesoamerica" (Inter-American Institute for Global Change Research); "Desarrollo sustentable de zonas cafetaleras" (Consejo Nacional de Ciencia y Tecnología, Mexico); "Fomento de la diversificación productiva en una organización de pequeños productores de café de Chiapas, México" (Peace Corps and ECOSUR, México); "Innovación socioambiental para el desarrollo en áreas de alta pobreza y biodiversidad de la frontera sur de México" (Fondo Institucional de Fomento Regional para el Desarrollo Científico, Tecnológico y de Innovación, Consejo Nacional de Ciencia y Tecnología, Mexico); "Small Coffee Producers and Adaptive Options for a Changing Climate: The Risks and Challenges of Certification for Ecosystem Services" (Inter-American Institute for Global Change Research); "Diseño, construcción, equipamiento y puesta en marcha de un Centro Estatal de Innovación y Transferencia de Tecnología para el Desarrollo de la Caficultura Chiapaneca" (Fondo Mixto Chiapas- Consejo Nacional de Ciencia y Tecnología, Mexico); and "Innovación socioambiental en zonas cafetaleras para la reducción de la vulnerabilidad" (ECOSUR, Mexico).

All these projects were carried out with the support of ECOSUR, the institution that has sheltered me for more than 35 years, and with which I am infinitely grateful.

I want to thank Pearson Education, Inc., New York, for allowing me to use Fig. 2.1; Island Press, Washington, DC for using Fig. 2.4; and Taylor and Francis Group, LLC, by using Figs. 5.4, 5.8, 6.2, 6.4, 6.5, 6.7, 6.8, 6.9, and 6.10, and Tables 7.1 and 7.2.

I also want to thank Dr. Aurelio Ciancio, Series Editor; Mr. Prasad Gurunadham, Project Co-ordinator (Books) for Springer Nature; and Mrs. Rathika Ramkumar,

Project Manager, Content Solutions, SPi Global, for their empathy and support during planning and editing this book.

I thank Jaime Gómez for his support in reviewing the first draft in Spanish of this book.

Finally, I am very grateful to Juan Gerardo Barrera Covarrubias for his wise comments to improve the first English version of this book.

Contents

About the Author

Juan F. Barrera (https://orcid.org/0000-0002-8488-7782) is a Researcher at El Colegio de la Frontera Sur (ECOSUR) in Tapachula, Chiapas, Mexico, since 1983.

He carried out his undergraduate studies as an Agricultural Engineer with a specialty in Parasitology at the Universidad Autónoma Agraria Antonio Narro (UAAAN, Saltillo, Mexico). He completed his master's degree in Agricultural Parasitology at the Instituto Tecnológico y Estudios Superiores de Monterrey (ITESM, Monterrey, Mexico) and obtained the title of Docteur from the Université Paul Sabatier (UPS, Toulouse, France) with specialty in Applied Entomology.

His research covers topics on ecology, biology, behavior, and pest sampling, as well as biological control and pest management in tropical crops, with an emphasis on the coffee berry borer.

He was President of the Mexican Society of Biological Control and is Member of the Mexican Academy of Sciences, the National System of Researchers of the National Council of Science and Technology of Mexico, and the Entomological Society of America.

For 15 years, he has been developing a new paradigm of plant protection with a holistic approach.

List of Figures

List of Tables

Chapter 1
Introduction: Beyond IPM

To be accepted as a paradigm, a theory must seem better than its competitors, but it doesn't have to explain all the facts it faces and in fact it never does.
Thomas S. Kuhn, The structure of scientific revolutions (Translated from: Kuhn TS (2006) La estructura de las revoluciones científicas. Fondo de Cultura Económica, 3rd ed, Mexico, p. 81.)

Abstract Integrated Pest Management (IPM), currently considered the paradigm or "normal science" of pest management, presents anomalies that show its exhaustion. Such anomalies are its low adoption rates by farmers, serious losses caused by pests despite advances in the science of crop protection, and the continuous emergence of alternative pest management strategies. The origin of these anomalies is attributed to the reductionist approach of IPM. Therefore, a drastic innovative change in pest management is proposed, based on a holistic approach. More specifically, it is necessary to develop a socio-environmental system where all pest management actions are interconnected with the other components of the system and are oriented to the people's well-being. In order to achieve this goal, Holistic Pest Management (HPM), a new paradigm of pest management, is proposed.

Keywords Integrated Pest Management · Holistic Pest Management · Paradigm · Low adoption rates · Farmers

J. F. Barrera, *Beyond IPM: Introduction to the Theory of Holistic Pest Management*, Sustainability in Plant and Crop Protection 15,
https://doi.org/10.1007/978-3-030-43370-3_1

1.1 Introduction

Throughout history, with the purpose of eliminating or counteracting the losses caused by pests,[1] the human being developed various technologies and implemented countless control programs, throughout the world.

Among the control strategies, the Integrated Pest Management (IPM) stands out. According to Peterson et al. (2018), the IPM has been an important success story for society in the fields of agriculture, applied ecology, urban management and public health. In many cases, the IPM has been shown to reduce populations and damage caused by pests at the lowest economic, social and environmental costs (Kogan 1998; Norris et al. 2003; Barzman et al. 2015; Farrar et al. 2015; Bottrell and Schoenly 2018).

Today few doubt that the IPM is not the most suitable strategy to deal with pests. In fact, IPM has been accepted and incorporated into the public policy of many countries (Stenberg 2017). Based on the concept that Kuhn (2006) proposes in his book "The structure of scientific revolutions", the IPM can be considered as the current paradigm or "normal science" of pest management.[2]

Following Kuhn's approach, a current paradigm begins to lose utility as "anomalies" arise, that is, unexplained situations under the arguments of that paradigm. At first, the anomalies are detected and proposed by a very small number of people and have no or very little impact on the perceptions and knowledge of the rest of the community. However, as time goes by, the anomalies become more evident and so does the number of people who take advantage of what, at any given time, will be accepted as the new paradigm, that will replace the current one and, with it, occurs what Kuhn calls a "revolution" of science.

The evolution of IPM[3] is a clear example of paradigm substitution under Kuhn's approach: first, what was called "Supervised Insect Control" emerged; then, the "Integrated Pest Control" was suggested; and later, the "Integrated Pest Management" was proposed, a term that prevails this day.

This chapter analyzes the main anomalies that, from our point of view, the IPM presents, anomalies that have led to the proposal of the Holistic Pest Management (HPM).

---

[1] From here on we will refer to a broad concept of "pest", indicating any organism that interferes with human activities and purposes.

[2] For a discussion of the Kuhn's paradigm concept applied to IPM, see Trumble (1998).

[3] Chapter 2 further analyzes the evolution of IPM.

1.2 Three Anomalies of the Current Paradigm

There are at least three facts or anomalies showing that, despite technological innovations and resources invested, the IPM has left much to be desired, especially in developing countries. These anomalies are:

- its adoption rates by farmers has been very low;
- crop losses to pests have not decreased globally; and
- the emergence of alternative strategies to IPM.

These anomalies are explained in the next sections.

1.2.1 Low Adoption Rates of IPM

Zalom (1993) recognized that IPM is far from being adopted by all farmers, even if the programs are available and validated.

IPM emerged as a strategy to reduce an excessive use of pesticides. However, its low adoption – particularly in developing countries (Bottrell and Schoenly 2018) – has not allowed a significant reduction of the use of these chemicals (Benbrook et al. 1996; Trumble 1998; Parsa et al. 2014; Jayasooriya and Aheeyar 2016; Carrión Yaguana et al. 2016; Alwang et al. 2019). For example, globally it is estimated that in the last 20 years the use of pesticides has increased to 3.5 billion kg/year, with a value around US $45 billion in the market (Pretty and Bharucha 2015).

The low adoption of IPM is multifactorial. According to Zalom (1993), there are technical, financial, educational, institutional and social restrictions for the use of IPM. Soon (1996) cites limitations for the adoption of IPM in developing countries such as *(i)* the perception that IPM is complex, *(ii)* public policies that favor the use of pesticides, *(iii)* the difficulties of joint work between researchers, extension agents and producers, *(iv)* the absence of biological control or disturbance of natural enemies by pesticides, *(v)* the lack of integration of different social actors, *(vi)* the lack of training or use of inappropriate training models and *(vii)* the lack of financing, poor financing or financing of non-priority areas.

The complexity of the IPM – as a factor that demands attention and resources to achieve its implementation – has been one of the main causes that affects its adoption (Wearing 1988). A more recent study based on interviews and workshops points out differences in the obstacles that limit the adoption of IPM between developed and developing countries (Parsa et al. 2014). In the case of Latin America, respondents felt that the most important obstacle was "IPM requires collective action within a farming community". Considering the people consulted in developed countries, the obstacle that occupied the first place was the "shortage of well-qualified IPM experts and extensionists".

In developed countries such as the USA and those that make up the European Union,[4] the low adoption of IPM by the small farms sector is due, according to Grasswitz (2019), to the fact that research and extension programs related to IPM mainly focus on large-scale agriculture.

1.2.2 Pest Losses

It is well known that pests are among the most important limiting factors in the productivity of agricultural, forestry and livestock systems; whether insects, pathogens or weeds, these organisms are responsible for large losses. The information on losses that these organisms cause is important both for their management and for evaluating the effectiveness of plant and animal protection practices (Oerke and Dehne 2004; Oliveira et al. 2014).

Worldwide, it is estimated that losses in crops caused by arthropods, diseases and weeds have increased from 34.9% in 1965 to 42.1% in 1988–1990 (Lewis et al. 1997).

An analysis of 17 regions for the 1996–1998 period estimated that potential losses by weeds, pest animals, pathogens and viruses and actual losses, that is, losses despite current crop protection practices, varied from less of 50% (in barley) to more than 80% (in sugar beet and cotton). With respect to real losses, they were estimated at 26–30% for sugar beets, barley, soybeans, wheat and cotton, and 35%, 39% and 40% for corn, potatoes and rice, respectively (Oerke and Dehne 2004).

Other analyses indicated that the overall potential losses by pests for the 2001–2003 period varied from 50% in wheat to more than 80% in cotton (Oerke 2006). The same source indicated that the losses had not decreased in the last 40 years despite the significant use of pesticides.

In Brazil, based on data from Goellner (1993) and Bento (1999) on losses caused by insect pests in 35 major crops and production and price data for 2012 and 2013, Oliveira et al. (2014) reported 2–43% of losses. Likewise, an average of 7.7% of losses caused by insects in these crops was estimated, which represented a reduction of approximately 25 million tons of food, fiber and biofuels. According to Oliveira et al. (2014), the estimated annual loss in money was approximately US $14.7 billion, despite the adoption of control measures.

It is surprising that, despite all the knowledge generated and all the technology developed, many pests continue to cause serious losses (Pimentel et al. 1991; Sweetmore et al. 2001; Oerke 2006; Culliney 2014; Reddy 2017). In some crops, losses have even remained roughly constant over time (Oerke 2006). In many cases,

[4]The Economic Research Service of the United States Department of Agriculture published on August 30, 2019 that just over half of all farms in the USA were very small, with annual agricultural sales of less than US $10,000 (USDA 2019); a survey conducted by Eurostat in 2016 showed that two thirds of all farms in the European Union were less than 5 hectares in size (Cook et al. 2018).

according to Altieri and Rosset (1996), crops losses were still caused by pests despite the excessive use of pesticides applied for their control, thus revealing the environmental crisis through which modern conventional agriculture passes.

1.2.3 Alternative Strategies to IPM

The first concept of IPM as actually known emerged 60 years ago (Stern et al. 1959). Since then, the concept of IPM has been constantly evolving (Bajwa and Kogan 1996, 2002), to the extent that some researchers have seen the need to propose changes in concept and approach. Since the 1980s there was a notable increase in the number of definitions of IPM, that continued to increase until the 1990s (Fig. 1.1).

Some examples of alternative proposals to IPM[5] are: "Biologically intensive IPM" (Frisbie and Smith 1991), "Agroecology and pest management" (Andow 1991; Altieri 1994), "Ecologically based pest management" (NRC 1996), "Total system approach" (Lewis et al. 1997), "Integrated biodiversity management" (Kiritani 2000), "Community MIP" (Fakih et al. 2003), "Integrated production" (Boller et al. 2004), and the "Crop health" (Vega et al. 2019), among others.

New proposals for the IPM concept that enhance or emphasize the importance of some aspects of this paradigm have also emerged. For example, Dara (2019) proposed a new model based on the ecological and economic bases of previous models.

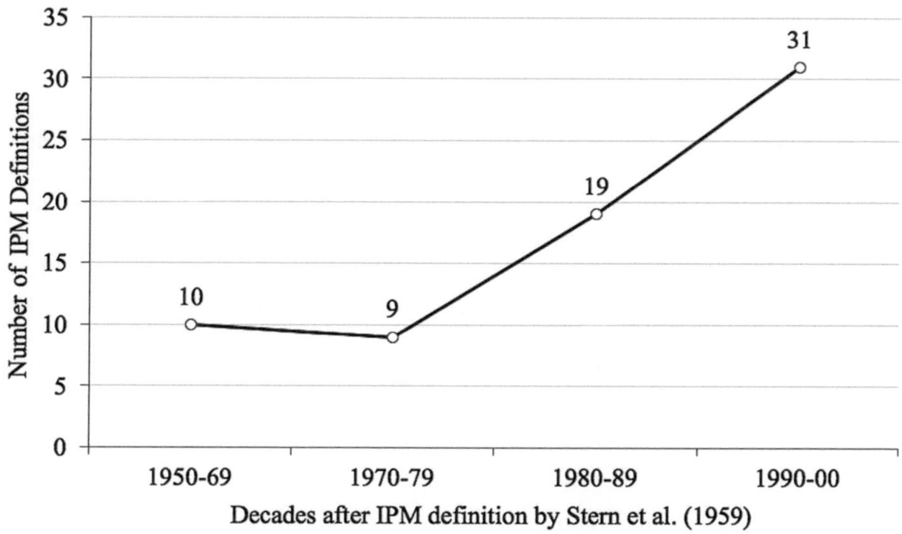

Fig. 1.1 Number of definitions of IPM per decade after the first definition published by Stern et al. (1959). Data taken from Bajwa and Kogan (1996, 2002)

[5]Chapter 2 describes the main characteristics of some alternative proposal to IPM.

To adjust it to modern times, Dara (2019) expanded and reconfigured the IPM concept to include management, business and sustainability aspects, emphasizing the importance of research and dissemination.

Such examples represent a sample of the conflicts derived from the concept, implementation and impact of IPM.

1.3 Beyond IPM

The low adoption of IPM by farmers, the severe losses that pests continue to cause despite many technological developments, and the continuous emergence of alternative strategies to IPM in different parts of the world, are nothing more than symptoms indicating that IPM –the current pest management paradigm– is exhausted.

If these are the symptoms of IPM depletion, what is or what are the causes? Our analysis of this problem led us to conclude that IPM anomalies are mainly due to the lack of a true holistic approach, that is to say, away from the rhetoric with which holism is treated in the definitions of IPM. To make progress in this direction, it is not only necessary to develop an holistic IPM approach, as proposed by Stenberg (2017), but it is necessary to go even further through an Holistic Pest Management, a new paradigm of pest management, that is proposed in this book (Barrera 2006, 2020).

1.4 Conclusion

The evolution of pest management to higher levels of knowledge, implementation and impact requires a drastic change of direction. It requires a system where every pest management action is interconnected with the other system components, in a socio-environmental system where pest management is geared towards people's well-being. More than 30 years ago Wearing (1988) pointed out that the main objective of an IPM program was to adjust to the needs of farmers in all aspects, whether social, economic, cultural, technical, organizational or political. For this, it is necessary to move from the IPM reductionist approach towards an holistic approach to pests. This book offers insight and ways to achieve it.

References

Altieri, M. A. (1994). *Biodiversity and pest management in agroecosystems*. New York: Hayworth Press.

Altieri, M. A., & Rosset, P. (1996). Agroecology and the conversion of lárge-scale conventional systems to sustainable management. *International Journal of Environmental Studies, 50*, 165–185.

Alwang, J., Norton, G., & Larochelle, C. (2019). Obstacles to widespread diffusion of IPM in developing countries: Lessons from the field. *Journal of Integrated Pest Management, 10*(1), 1–8.

Andow, D. A. (1991). Vegetational diversity and arthropod population response. *Annual Review of Entomology, 35*, 561–568.

Bajwa, W. I., & Kogan, M. (1996). *Compendium of IPM definitions (CID).* http://www.ipmnet.org/ipmdefinitions/defineIII.html#90's. Accessed 14 Jul 2019.

Bajwa, W. I., & Kogan, M. (2002). *Compendium of IPM definitions (CID) – What is IPM and how is it defined in the Worldwide Literature?* IPPC Publication No. 998, Integrated Plant Protection Center (IPPC), Oregon State University, Corvallis, Oregon.

Barrera, J. F. (2006). Manejo holístico de plagas: Hacia un nuevo paradigma de la protección fitosanitaria. In J. Pohlan, L. Soto, & J. Barrera (Eds.), *El cafetal del futuro: Realidades y Visiones* (pp. 63–82). Aachen: Shaker Verlag.

Barrera, J. F. (2020). Holistic Pest Management. In D. Perez-Staples, F. Diaz-Fleischer, P. Montoya, & M.T. Vera (Eds.), Area-wide management of fruit fly pests (pp. 275–293). Boca Raton: CRC Press, Taylor & Francis Group, LLC.

Barzman, M., Bàrberi, P., Birch, A. N. E., et al. (2015). Eight principles of integrated pest management. *Agronomy for Sustainable Development, 35*(4), 1199–1215.

Benbrook, C. M., Groth, E., III, Halloran, J. M., et al. (1996). *Pest management at the crossroads.* Yonkers: Consumers Union of United States.

Bento, J. M. S. (1999). Perdas por insetos na agricultura. *Ação Ambient, 4*, 19–21.

Boller, E. F., Avilla, J., Joerg, E., et al. (Eds.). (2004). Integrated production. Principles and technical guidelines. 3rd Edition. *IOBC WPRS Bulletin, 27*(2), 1–49.

Bottrell, D. G., & Schoenly, K. G. (2018). Integrated pest management for resource-limited farmers: Challenges for achieving ecological, social and economic sustainability. *Journal of Agriculture Science, 156*, 408–426.

Carrión Yaguana, V., Alwang, J., Norton, G., et al. (2016). Does IPM have staying power? Revisiting a potato-producing area years after formal training ended. *Journal of Agricultural Economics, 67*(2), 308–323.

Cook, E., Rybkowska, A., & Selenius, J. (2018). Farm structure survey 2016. *Eurostat Newsrelease.* 105/2018-28 June 2018.

Culliney, T. W. (2014). Crop losses to arthropods. In D. Pimentel & R. Peshin (Eds.), *Integrated pest management* (pp. 201–225). Dordrecht: Springer.

Dara, S. K. (2019). The new integrated pest management paradigm for the Modern Age. *Journal of Integrated Pest Management, 10*(1), 12, 1–9.

Fakih, M., Rahardjo, T., & Pimbert, M. P. (2003). *Community integrated pest management in Indonesia: International Institute for Environment and Development.* Brighton: Institute for Development Studies.

Farrar, J., Baur, M., & Elliott, S. (2015). *Adoption and impacts of integrated pest management in agriculture in the western United States.* Davis, California: Western IPM Center.

Frisbie, R. E., & Smith Jr, J. W. (1991). Biologically intensive integrated pest management: The future. In J. J. Menn & A. L. Steinhauer (Eds.), *Progress and perspectives for the 21st century.* Entomological Society of America, Centennial National Symposium. Lanham, MD (pp. 151–164).

Goellner, C. I. (1993). *Utilização dos defensivos agrícolas no Brasil: Análise do seu impacto sobre o ambiente e a saúde humana.* Artgraph, São Paulo, Brazil

Grasswitz, T. R. (2019). Integrated Pest Management (IPM) for small-scale farms in developed economies: Challenges and opportunities. *Insects, 10*, 179.

Jayasooriya, H. J. C., & Aheeyar, M. M. M. (2016). Adoption and factors affecting on adoption of integrated pest management among vegetable farmers in Sri Lanka. *Procedia Food Science, 6*, 208–212.

Kiritani, K. (2000). Integrated biodiversity management in paddy fields: Shift of paradigm from IPM toward IBM. *Integrated Pest Management Reviews, 5*(3), 175–183.

Kogan, M. (1998). Integrated pest management: Historical perspectives and contemporary developments. *Annual Review of Entomology, 43*, 243–270.

Kuhn, T. S. (2006). *La estructura de las revoluciones científicas* (3ª ed). Mexico: Fondo de Cultura Económica.

Lewis, W. J., van Lenteren, J. C., Phatak, S. C., et al. (1997). A total system approach to sustainable pest management. *Proceedings of the National Academy of Sciences of the United States of America, 94*, 12243–12248.

NRC (National Research Council). (1996). *Ecologically based pest management: New solutions for a new century*. Washington, DC: National Academy Press.

Norris, R. F., Caswell-Chen, E. P., & Kogan, M. (2003). *Concepts in integrated pest management*. Upper Saddle River: Prentice Hall.

Oerke, E.-C. (2006). Crop losses to pests. *Journal of Agriculture Science (Cambridge), 144*(1), 31–43.

Oerke, E.-C., & Dehne, H. W. (2004). Safeguarding production- losses in major crops and the role of crop protection. *Crop Protection, 23*(4), 275–285.

Oliveira, C. M., Auad, A. M., Mendes, S. M., et al. (2014). Crop losses and the economic impact of insect pests on Brazilian agriculture. *Crop Protection, 56*, 50–54.

Parsa, S., Morse, S., Bonifacio, A., et al. (2014). Obstacles to integrated pest management adoption in developing countries. *Proceedings of the National Academy of Sciences of the United States of America, 111*(10), 3889–3894.

Peterson, R. K. D., Higley, L. G., & Pedigo, L. P. (2018). Whatever happened to IPM? *American Entomologist, 64*(3), 146–150.

Pimentel, D., McLaughlin, L., Zepp, A., et al. (1991). Environmental and economic effects of reducing pesticide use. *BioScience, 41*(6), 402–409.

Pretty, J., & Bharucha, Z. P. (2015). Integrated Pest Management for sustainable intensification of agriculture in Asia and Africa. *Insects, 6*(1), 152–182.

Reddy, P. P. (2017). *Agro-ecological approaches to pest management for sustainable agriculture*. Singapore: Springer.

Soon, L. G. (1996). Integrated pest management in developing countries. In G. J. Persley (Ed.), *Biotechnology and integrated pest management* (pp. 61–75). Wallingford: CAB International.

Stenberg, J. A. (2017). A conceptual framework for integrated pest management. *Trends in Plant Science, 22*(9), 759–769.

Stern, V. M., Smith, R. F., van den Bosch, R., et al. (1959). The integrated control concept. *Hilgardia, 29*(2), 81–101.

Sweetmore, A., Rothschild, G., & Eden-Green, S. (Eds.). (2001). *Perspectives on pests: Achievements of research under the UK Department for International Development's Crop Protection Programme, 1996-2000*. National Resource International Limited, UK.

Trumble, J. T. (1998). IPM: overcoming conflicts in adoption. *Integrated Pest Management Reviews, 3*(4), 195–207.

USDA (United States Department of Agriculture). (2019). Farming and farm income. *Economic Research Service*. https://www.ers.usda.gov/data-products/ag-and-food-statistics-charting-the-essentials/farming-and-farmincome/ Accessed 18 Nov 2019.

Vega, D., Gazzano Santos, M. I., Salas-Zapata, W., et al. (2019). Revising the concept of crop health from an agroecological perspective. *Agroecology and Sustainable Food Systems*. Published online: 24 Jul 2019.

Wearing, C. H. (1988). Evaluating the IPM implementation process. *Annual Review of Entomology, 33*, 17–38.

Zalom, F. G. (1993). Reorganizing to facilitate the development and use of integrated pest management. *Agriculture, Ecosystems & Environment, 46*(1–4), 245–256.

Chapter 2
The Nature of Integrated Pest Management

"Nothing is born from nothing"
Tito Lucrezio Caro, The Nature of Things, Book First, Row 211 (Biblioteca Virtual Universal. 2003. Editorial del Cardo https://www.biblioteca.org.ar/libros/89401.pdf)

Abstract This chapter aims to find out the complexities of the nature of Integrated Pest Management (IPM) and to present some points made by its critics. It begins by presenting the evolution of IPM: first, the Supervised Insect Control is presented, followed by the presentation of the Integrated Pest Control and, finally, exposing what is known today as IPM. Next, an analysis is made of the almost 70 definitions of IPM that have emerged over time. Then, the long-standing relationship of IPM with pesticides and their impacts is explored. It continues with a section that presents some criticism to IPM. These indicate that the nature of IPM encourages the substitution of inputs, the search for the "silver bullet", justifying the use of pesticides through thresholds of action. Some alternative strategies to IPM that have emerged in the last 30 years are reviewed. Finally, the adaptive renewal cycle of IPM is presented.

Keywords Holistic Pest Management · IPM definitions · Criticism to IPM · Pesticides · Adaptive renewal cycle

2.1 Introduction

IPM was proposed as a response to excessive use of pesticides and to reduce the harmful effects of these chemicals on the environment and human health (Altieri 2018). However, the increase in pesticide imports by developing countries (Bottrell and Schoenly 2018), suggests that these products continue to be overused.

J. F. Barrera, *Beyond IPM: Introduction to the Theory of Holistic Pest Management*, Sustainability in Plant and Crop Protection 15,
https://doi.org/10.1007/978-3-030-43370-3_2

Understanding the nature of IPM was the approach we took to review its "anomalies" and propose the Holistic Pest Management (MHP) as an alternative paradigm. In this sense, knowing the historical evolution of the IPM concept, its multiple definitions and its long-standing relationship with pesticides, are key elements for the analysis.

This chapter aims to find out the complexities of the nature of IPM and expose some of its criticism. Also, some alternative strategies to IPM that have emerged in the last 30 years are reviewed.

2.2 Evolution of IPM

Humans have been in direct competition with thousands of pests since immemorial time (Gray et al. 2009). Since then, pest control has been required to protect the production and quality of food demanded by the human population (Popp et al. 2013).

Evolutionarily speaking, IPM has not been a static concept, but a succession of three clearly defined paradigms that began in the 1950s in USA with the emergence of organosynthetic pesticides (Kogan 1998; Ehler and Bottrell 2000):

(i) Supervised Insect Control
(ii) Integrated Pest Control
(iii) Integrated Pest Management

Each of these paradigms emerged in response to "anomalies" observed by some scholars and that were accepted by the community at the time.

2.2.1 Supervised Insect Control

Shortly after World War II –from the late 1940s until the early 1950s– when synthetic insecticides were available, entomologists from California and Arkansas (USA) developed the paradigm of "Supervised Insect Control" (SIC) when working with cotton crop pests (Ehler and Bottrell 2000). In SIC, qualified entomologists "supervised" the control and applications of insecticides based on monitoring of pests and their natural enemies, instead of using the scheduled applications that were a standard practice. The SIC contributed in such an important way so that, in the 1950s, California entomologists developed the conceptual basis of "Integrated Pest Control" (IPC).

2.2.2 *Integrated Pest Control*

In the 1960s, interest in IPC grew due to concerns about the excessive use of pesticides in agriculture and their harmful effects (Corbet and Smith 1975). Initially, the IPC had the purpose of integrating chemical control with biological control in order to counteract the undesirable effects of abuse and misuse of synthetic pesticides. In this regard, Vernon M. Stern (1923–2006), Ray F. Smith (1919–1999), Robert van den Bosch (1922–1978) and Kenneth S. Hagen (1919–1997) published in 1959, a seminal paper where they established the principles and philosophy of IPC and proposed the concepts of "Economic Injury Level" (EIL) and "Economic Threshold" (ET) (Stern et al. 1959), considered the cornerstone of IPM.

It is worth mentioning that in 1981 the paper of Stern et al. (1959) was acclaimed by the US National Academy of Sciences as "the single most important paper published on crop protection in this century", when one of its authors, Dr. Ray F. Smith, was elected member by the academy (Warnert 2009). Even today, the ecological principles mentioned in the Stern et al. (1959) paper are considered as valid as they were more than 60 years ago (Bottrell and Schoenly 2018). For their remarkable contribution, Stern, Smith, van den Bosch and Hagen are considered the founders of IPM (Warnert 2009).

Defined as a pest management system that uses all suitable techniques to reduce and maintain pest populations at levels below those that cause economic injury, or manipulate these populations to prevent injury (van den Bosch et al. 1971), IPC represented at the beginning two basic ideas: consider the ecosystem and use action thresholds (Corbet and Smith 1975).

2.2.3 *Integrated Pest Management*

The term "control" was questioned by some researchers since the origins of the IPC (Geier 1966; Smith and Reynolds 1966); instead, they proposed the term "management", resulting in "Integrated Pest Management" or IPM.

Higley and Pedigo (1993) pointed out that there are important differences between control and management. While control refers to having an influence on something, for example, reducing the number of pests to a desired degree (generally to zero), management refers to the judicious use of the means to achieve a desired end, that is, management focuses on reducing or modifying the impact of pests to tolerable levels.

The definition of IPM provided by these researchers was:

> "Integrated pest control is a pest population management system that utilizes all suitable techniques in a compatible manner to reduce pest populations and maintain them at levels below those causing economic injury" (Geier 1966; Smith and Reynolds 1966).

Bottrell (1979) was one of the pioneer researchers who explicitly established the concept of IPM and stressed the importance of the following principles:

- Potentially harmful species continue to exist at tolerable levels of abundance.
- The ecosystem is the management unit.
- The use of natural enemies is maximized.
- Any control method can produce unexpected and undesirable effects.
- An interdisciplinary approach is essential.

However, since the early 1980s an analysis of several IPM programs in the USA carried out by Barfield and O'Neil (1984) revealed that these programs had the following characteristics:

(i) There was a tendency to consider the crop field as the limit of the system, rather than the ecosystem as a whole.
(ii) There was little understanding (if any) of the mortality exerted by natural enemies.
(iii) The pests were treated by group (insects, weeds, pathogens, nematodes), instead of being seen in an integrated way.
(iv) The action thresholds were considered static, not as a function of changes in the managed system.
(v) Pest population monitoring was frequently inaccurate or absent.

Today, IPM is considered a sophisticated system that attempts to ensure the sustainability of an agricultural system, as well as the conservation of its environment (Kogan 1998). One of the most recent definitions of IPM says:

> "Integrated Pest Management is a decision-making assistance system to select and use pest control tactics alone or coordinated in harmony within a management strategy, based on cost-benefit analysis that takes into account the interests of and the impacts on producers, society and environment" (Norris et al. 2003) (Fig. 2.1).

However, an irrefutable fact is that the practice of IPM has not always been consistent with the theory (NRC 1996; Altieri 2018). To understand this paradox, the relationship between IPM and pesticides, an old-fashioned relationship, will be analyzed first. Later, some criticisms to IPM will be presented; as we will see, these criticisms have arisen following the reductionist approach of IPM.

2.3 Analysis of IPM Definitions

Over time, the concept of IPM evolved. According to Bajwa and Kogan (1996, 2002), more than 60 definitions related to the IPM concept were coined between 1950 and 2000. When analyzing the frequency in which the most important terms and expressions occur in the definitions of IPM presented by these authors (Fig. 2.2), it is observed that expressions that use "economic" and "environmental" terms predominate, with more than 48%; and terms such as "pest populations" and "pest control" follow, with more than 38% of the mentions. Terms like "system" and

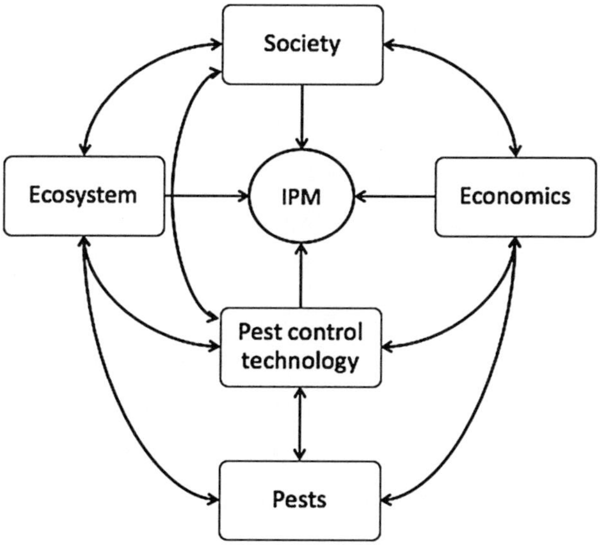

Fig. 2.1 Conceptual diagram of Integrated Pest Management (IPM) showing its components and relationships. If pests (base of the diagram) do not exist, pest management is not necessary. Based on understanding the ecology and biology of pest populations, the basis for developing the management program is established. The management of pest populations includes the use of technologies based on cultural, mechanical, physical, biological, genetic or chemical control methods. Decision making to use these methods depends on economic and environmental factors. The opinions and limitations imposed by society (at the top of the diagram) affect all aspects of IPM. (Based on Norris et al. 2003). © 2003 From Concepts in Integrated Pest Management by Norris R.F., Caswell-Chen E.P., & Kogan, M. Reproduced by permission of Pearson Education, Inc., New York, NY

"social and/or sociological" aspects represented a lower frequency with 24.2% and 9.7%, respectively.

Using a word cloud generator software[1] (Fig. 2.3) to analyze the definitions of IPM compiled by Bajwa and Kogan (1996, 2002), in groups for periods of approximately 10 years after the IPM definition proposed by Stern et al. (1959), we found the following:

- The frequency of the word "management", comparatively low in the first years (1950–1969), increases markedly after this period; the opposite happens with the word "control", although it does not lose much presence in IPM definitions.
- "Integration", a keyword in IPM definitions, maintains a constant presence over time; something similar happens with the words "system" and "pest".
- The word "environment" starts to stand out as of the 1970s and over time it becomes important in the definitions of IPM. Related words such as "ecological" or "ecologically" are present (with the exception of the 1980s), although less frequently than "environment".

[1] Word cloud generator software: https://www.wordclouds.com

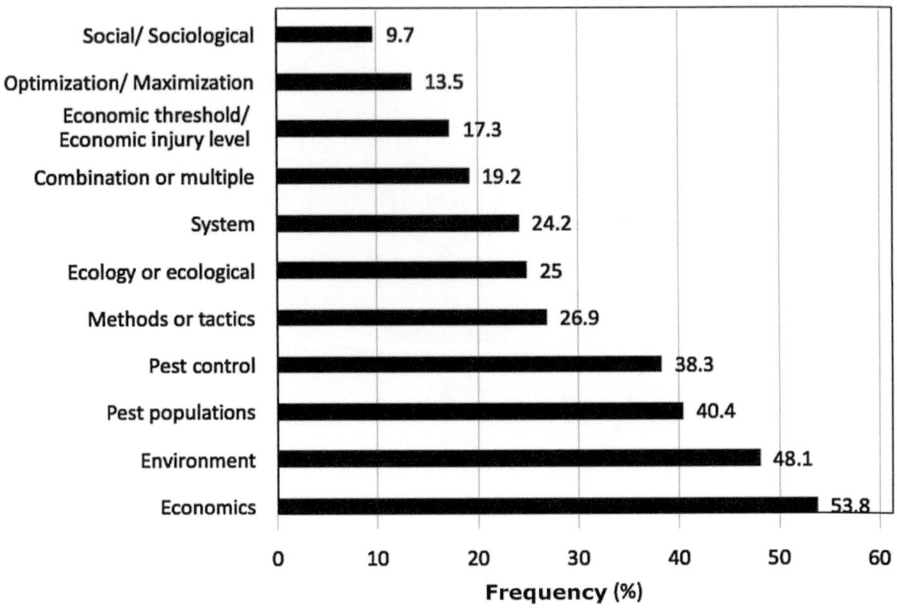

Fig. 2.2 Frequency of terms or expressions used in 67 definitions of IPM compiled in the Compendium of IPM definitions by Bajwa and Kogan (1996, 2002). (Data from Bajwa and Kogan 2002)

- The word "chemical", related to chemical control, relatively present in IPM definitions in the first period, lost presence in the 1970s but later resurfaced, with an important presence in the 1990s; the word "pesticides" is also noted in this last period.
- The frequency of "biological", which refers to biological control, is clearly defined in the first (1950–1969) and last (1990–2000) periods, while losing presence in the 1970s and 1980s; in these, biological control is represented by words related to natural enemies ("natural" and "enemies") or "parasites" and "predators".
- "Cultural", of cultural control, one important method of IPM, appears with forcefulness (almost like the word "chemical") in the definitions of the 1990s whereas in the previous decade it barely figures.
- With the exception of the 1980s, the word "economic" is well represented in the definitions of IPM.
- "Population" and "populations" are present in IPM definitions in all periods except the last one (1990–2000), where they barely appear; something similar happens with action thresholds ("level", "levels", "action").
- Some words like "weather" and "production" appeared in IPM definitions only in the 1980–1989 period.

Fig. 2.3 Word clouds elaborated with definitions of IPM concept over time. The more the words are repeated in the analyzed text, the larger their size . The IPM definitions were taken from Bajwa and Kogan (1996, 2002). (**a**) 1950–1969; (**b**) 1970–1979; (**c**) 1980–1989; (**d**) 1990–2000.

- From the 1970s, the word "crop" is present in IPM definitions, although with a moderate frequency.
- In the 1990s words with little or no presence in previous periods emerge in IPM definitions, such as "health", "public", "human", "society", "risks" and "information".
- Words like "ecosystem" or "interdisciplinary" mentioned in the principles of Bottrell (1979) do not appear among the most cited words in IPM definitions. Nor do terms that stand out today in vogue as "sustainable", "food safety" or "farmer participation".
- The word "holistic" goes unnoticed in the definitions of IPM over the time considered in this analysis.

The analysis of the most frequent words in IPM definitions is more than eloquent about its nature. In fact, it tells us about an IPM concerned about the environment but with chemical control as favorite methods; an IPM with a solid economic base but neglecting social aspects, although in recent times these aspects have become more important; an IPM whose universe of attention is limited to the crop and the pest, leaving aside other levels of integration; in short, an IPM that has favored the reductionist approach.

2.4 IPM and Pesticides

Undoubtedly, the use of pesticides[2] has given humanity great benefits in pest control. For example, insecticides contributed to reduction – even disappearance – of several diseases in entire regions by eliminating arthropod vectors of malaria. Also, the use of pesticides as part of technological packages during the Green Revolution was essential to bring food production to levels never seen before, contributing to improved varieties expressing their maximum genetic potential (Andrews 1989; Kogan 1998). Globally, it has been estimated that farmers use around USD $40 billion in pesticides per year (Popp et al. 2013).

A review conducted in the 1970s concluded that most cases of pest control used only pesticide application (Corbet and Smith 1975). Paradoxically, the indiscriminate and extensive use of pesticides in many places meant that crops such as cotton, citrus and rice were not profitable by increasing production costs without decreasing the losses caused by pests. In addition, the excessive use of pesticides aroused, as never before, the concern of society for damage to health, the impact on biodiversity and environmental pollution (Hollingsworth and Coli 2004; Aktar et al. 2009).

In developing countries, many traditional agricultural practices were replaced by pesticides (Dreves 1996). One of the most notable effects of repeated applications of the same pesticide is the selection of pest resistant populations, resulting in farmers having to apply more and more pesticides to achieve the same effect (Metcalf 1989; Hoy 2009). Less notable, but not less important, has been the elimination of beneficial fauna that contributes to natural regulation of phytophagous populations. Thus, without regulation exercised by natural enemies, the target pests resurface more strongly; in addition, organisms that were not previously known as pests appear now causing damage (secondary pests). The loss of self-defense capacity of agroecosystems through natural biological control caused by pesticides (high vulnerability to pest outbreaks and low resilience to pests), pushed entire systems into a vortex or vicious circle ("pesticide treadmill") that ultimately led to their economic, social and environmental collapse (Bottrell 1979; Barfield and O'Neil 1984; Metcalf 1989; Cate and Hinkle 1993; Morse and Buhler 1997; Norris et al. 2003).

Rachel Carson (1907–1964) contributed in a very significant way to developing society and government's concern for the environment through her book "Silent Spring" published in 1962. In her book, she warned about the indiscriminate use of DDT and other pesticides (Carson 1980). The pesticide industry furiously attacked Dr. Carson as hysterical and alarmist, however, her complaint – now considered as the origin of the environmental movement – forced the USA Government to legislate on use of pesticides (Graham 1970; Anonymous 2007).

---

[2]According to the U.S. Environmental Protection Agency, a pesticide is defined as any substance or mixture of substances intended to prevent, destroy, repel or mitigate any pest. These include insecticides, herbicides, fungicides and other substances used to control pests such as plant regulators, defoliants or desiccants. Most pesticides have some risk of harm to humans, animals or the environment, however, biological pesticides (pheromones, microbial pesticides), are safer than traditional chemical ones (Olle 2000).

In order to reverse the disaster in which agroecosystems managed with excessive use of pesticides, the concept and method of IPM originated and developed. To leave the quagmire, the use of other control methods such as cultural and biological methods was privileged through IPM, leaving pesticides as the "last line of defense" and using them only after field monitoring justified their use. It can be concluded, then, that overuse and misuse of pesticides gave birth to IPM. Under this argument, Morse and Buhler (1997) pointed out that IPM is a "reactive" philosophy, not a "proactive" one. According to them, without the problems caused by pesticides, the IPM would not have become what it is today; in fact, the main motivation to resort to IPM, according to Corbet and Smith (1975), is when the use of chemical control is not practical.

2.5 Criticisms to IPM

According to Ehler and Bottrell (2000), much of what is known as "modern IPM" has been nothing other than the reinvention of Supervised Insect Control. At that time, as now, technical personnel monitored the crops and determined the most appropriate time to deal with a pesticide, this being frequently the main or only tactic used.

In the same vein, Altieri and Nicholls (2003) pointed out that today's IPM has not addressed the ecological causes of the environmental problems of modern agriculture, because the narrow vision of finding specific causes that affect productivity still prevails.

The following are three criticisms to IPM:

(i) Substitution of supplies.
(ii) Law of the minimum.
(iii) Action thresholds.

2.5.1 Substitution of Supplies

Many IPM projects and agrochemical companies have placed emphasis on replacing conventional pesticides (e.g., organochlorine insecticides, carbamates or organophosphates) with biological products such as *Bacillus thuringiensis* – a microbial pesticide – or massive releases of parasitoid or predatory insects. Even, genetically modified crops (e.g., Bt corn) and countless "green or soft products" – which in recent times have flooded the market – have been promoted with the same principle. The replacement of dangerous pesticides with those that predict less undesirable effects on the environment and human health is called "substitution of supplies" (Altieri et al. 1997; Altieri and Nicholls 2003). Basically, the substitution of supplies drives the use of pesticides as a control method and favors monoculture as a way of land use and cultivation.

2.5.2 *Law of the Minimum*

Coming from Agronomy, the "Law of the Minimum" is a dogma that we own to the German chemist Justus von Liebig (1803–1873). This law is based on the idea that the growth of an individual or population is limited by the essential nutrient found in smaller quantities than needed (Smith and Smith 2001). Therefore, the limit of the limiting factor could only be exceeded or corrected with an appropriate external input. Paraphrasing Liebig, nitrogen deficiency (limiting factor) in a crop can be corrected by applying urea (external supply or input), a fertilizer that will increase production; by correcting the nitrogen deficiency, another factor, for example an insect pest, would in turn become a limiting factor by increasing its populations in response to nitrogen levels contained in the crop foliage. Thus, the control of insect pest will require a new input (insecticide) to avoid crop losses. In this way, in IPM a process based on treating the symptoms is more perpetuated than the real causes that cause the ecological imbalance (Altieri and Rosset 1996; Dreves 1996; Altieri and Nicholls 2003). The conception that emanates from the Law of the Minimum is closely related to the so-called "silver bullet", that is, the technology that emphasizes finding and using the most powerful means to kill pests (Pedigo and Higley 1996).

2.5.3 *Action Thresholds*

Action thresholds – Economic Injury Level and Economic Threshold – are considered core concepts of IPM. These thresholds arose from the idea of tolerating pests as long as they do not cause harm; this involved determining the status of pests as the basis for prudent use of pesticides (Higley and Pedigo 1993; Pedigo and Higley 1996).

The Economic Injury Level (EIL) is calculated with the following equation (Higley and Pedigo 1996; Pedigo 1999):

$$\mathrm{EIL} = \mathrm{C} / \mathrm{VDIK}$$

where C = Cost of control; V = Value of the product in the market; I = Damage units/insect/yield unit; D = Loss per unit of injury; and K = Proportional reduction of injury due to control. For its part, the Economic Threshold (EU) is a threshold that is set below the EIL to warn of the need to take some control measure (usually the application of a pesticide).

The concepts of action thresholds, first proposed by Stern et al. (1959), imply that no control action should be carried out until the pest population reaches the level at which economic losses could occur. Under these concepts the use of pesticides is avoided following preset schedules and, consequently, the cost of unnecessary applications is saved and the cost of environmental impact is reduced.

However, it is criticized that in practice the pesticide industry has found in action thresholds the means to justify the use of pesticides, to the extent that many IPM programs are mainly pesticide management programs (Morse and Buhler 1997). Another criticism suggests that the concept of action thresholds is so simple that it ignores the influence of other factors, both inside and outside the crop-pest system (Pedigo and Higley 1996).

2.6 Some Alternative Approaches to IPM

The path towards sustainable development, which is increasingly necessary, poses great challenges for the survival of an IPM based on a reductionist approach. Therefore, there are no few alternatives to IPM that have emerged in the last 30 years, ranging from "Biologically Intensive IPM" (Frisbie and Smith 1991) to pest management with a total system approach (Lewis et al. 1997), through "IPM with agroecological approach" (Altieri 1994) and "Ecologically based pest management" (NRC 1996). Other proposals suggest a more comprehensive approach to pest management with natural resources such as "Integrated Biodiversity Management" (Kiritani 2000), "Integrated Crops Management" (Bradley et al. 2002), "Integrated Production" (Boller et al. 2004) or "Crop health" (Vega et al. 2019). Next, a sample of some of these approaches will be briefly reviewed.

2.6.1 Biologically Intensive IPM

Frisbie and Smith (1991) proposed the term "Biologically intensive IPM" or "Biointensive IPM". According to Benbrook et al. (1996), the biologically intensive IPM follows a systems approach to pest management based on an understanding of the pest ecology. This approach begins with an accurate diagnosis of the nature and causes of the problems that result in the appearance of pests, in order to propose preventive and biological methods that keep their populations within acceptable limits. It is based on resistant varieties and promotes plant health, crop rotation, disruption of pest reproduction and management of biological processes to diversify and increase populations of natural enemies. If these tactics are not effective, low-risk pesticides (e.g., biopesticides) are used as a last resort and carefully to minimize risks. A central characteristic of biointensive IPM is its dependence on information as a key input for management decisions making with opportunity.

2.6.2 Agroecology and Pest Management

Andow (1991) presented evidence of the importance of biodiversity to improve pest management, and based on that, Altieri (1994) analyzed the concept of designing and building vegetation architectures to protect populations of natural enemies or with direct deterrent effects on pest herbivores. This approach emphasizes the restoration of natural control mechanisms by adding a diversity of plants, rather than forcing the establishment of biological control in situations where the essential ecological elements are lacking (e.g. monocultures). Aim is to achieve optimal performance of natural enemies. Vegetation management not only regulates pests, but also conserves energy, improves soil fertility, minimizes erosion risks and reduces dependence on external inputs.

2.6.3 Ecologically Based Pest Management

Recommended as a cost-effective, safe and durable pest control approach, Ecologically Based Pest Management (EBPM) was proposed by a committee established by the US National Research Council (NRC 1996). It is based first on the biological knowledge of pests, and secondarily on control methods. This approach is built on the understanding of the managed system, including the natural processes of pest suppression. In contrast to common agricultural practices that cause disturbance and destabilize natural processes of pest suppression, this proposal recommends improving natural processes through practices complemented with biological control products and organisms, resistant plants and pesticides with a reduced spectrum of action. The EBPM is based on cultural and biological approaches to pest management, but takes advantage of technological advances to improve basic knowledge and, if necessary, uses chemical, biological or physical inputs to achieve the objectives of profitability, safety and durability. A variant of this alternative was proposed by Tshernyshev (1995), who mentions that this paradigm – proposed to replace IPM in the future – uses many elements of the IPM, but is based on maintaining the stability of the agroecosystem and suppressing pest outbreaks at the very beginning, when their populations are still low.

2.6.4 Total System Approach

In a call qualified as urgent, Lewis et al. (1997) proposed a fundamental shift towards the total system approach in crop protection to resolve the escalation of economic and environmental problems as a result of the fight against agricultural pests. In their proposal, they pointed out that pest management strategies have been dominated by the search for "silver bullet" type products, for control of infestation outbreaks. They also indicated that the management of undesirable variables in the ecosystem should be similar to that used in other systems such as the human body or the social order. Experience in these fields, they pointed out, support the fact that therapeutic interventions in any system are effective only for short periods, since these externalities are soon "neutralized" by counterattacks of the same system. According to the total system approach, long-term solutions can only be achieved by restructuring and managing the systems in such a way that they maximize the order of internal preventive forces. In other words, therapeutic tactics are required that reinforce natural regulators such as the inherent plants' defenses, crop mixtures, soil and natural enemies, among other system components. Thus, the central foundation should be an approach that takes into account ecosystem interactions and seeks solutions that have net benefits at level of the entire ecosystem. Pesticides and other approaches that treat symptoms, not being sustainable, should only be used as a last resort. Under this principle, Lewis et al. (1997) proposed a management strategy that begins with the question: "Why is the pest a pest?" and try to find the weak points of the agroecosystem that have led to these organisms becoming pests.

2.6.5 Integrated Management of Biodiversity

The "Integrated Management of Biodiversity" (IMB) is an operational concept of agroecology that reconciles IPM and biodiversity conservation, making them compatible each other (Kiritani 2000). While the densities of the pest species remain below the EIL, the species that are to be conserved must be managed to keep them above the extinction threshold. To be implemented, the IMB requires that there be sufficient habitats in the range of distribution of the species that are to be conserved, so that they complete their life cycles and their populations persist.

2.6.6 Integrated Crop Management

Integrated Crop Management (ICM) proposes to integrate the management of individual crops to benefit from interactions between them, such as pest control or maintaining soil fertility (Bradley et al. 2002). The integration of crop production strategies is an ancient technique that under the ICM approach takes advantage of modern technology to improve the system. Three components, in order of importance, characterize the different concepts of the ICM: environmental aspects, economic viability (food production must be seen as a business) and modern technologies (in organic agriculture these techniques are seen as artificial inputs). ICM seeks the efficient use of inputs and considers crop rotation as a long-term strategy to minimize weed and pest problems. In summary, ICM is an environmentally sensitive and economically viable production system or process that uses the latest available techniques to produce high quality food efficiently.

2.6.7 Integrated Production

The concept of "Integrated Production" was created under the leadership of the International Organization of Biological Control (IOCB) (Boller et al. 2004). This approach consists in producing high quality food and other agricultural products through the use of natural resources and regulatory mechanisms, to replace polluting inputs and ensure the system sustainability . Integrated Production emphasizes a holistic approach to systems that involves every farm as the basic unit. Other components of Integrated Production are: the central function of the agroecosystem, balanced nutrient cycles and the welfare of all animal species in breeding. Likewise, conservation and improvement of soil fertility, a diversified environment and the observance of ethical and social criteria are essential components. In Integrated Production the biological, technical and chemical methods are carefully balanced taking into account environmental protection, profitability and social requirements.

2.6.8 Crop Health

Vega et al. (2019) redefined the concept of "crop health" from a holistic, transdisciplinary and agroecological perspective. They define a healthy farming system as "*one that tends to achieve short-, medium-, and long-term goals of the farmer (or community of farmers), contributing to meeting the needs of society, enhancing ecological and social processes by minimizing reliance on external inputs and maximizing safety for humans and ecosystems*". Since the main objective is to promote crop health, they propose to work with people: value the local knowledge of farmers, interact in a collaborative social learning process and seek means to strengthen the natural resource base, and the collective organization of farmers. These authors propose four components to assess the crop health status: usefulness, adversities, safety and autonomy. In addition, three components of health promotion from the proposed "salutogenic" model for human health (Antonovsky 1996), were adapted to the crop health concept: meaningfulness (i.e., that the challenge imposed makes sense to the person), comprehensibility (the cognitive aspect), and manageability (what the person does or can do in response to a challenge).

2.7 The Adaptive Renewal Cycle of Pest Management

The adaptive cycle is a fundamental unit for understanding complex systems, from cells to ecosystems, or societies to cultures, that can help to understand pest management responses to crises. The adaptive cycle or adaptive renewal cycle is a concept proposed by Holling (2001) and Holling and Gunderson (2002) in reference to changes that economic, ecological and social systems undergo when facing crises. The shape of adaptive cycle and the future state of the system depend on three properties (Holling 2001):

(i) The richness or inherent potential of a system that is available for change. This property sets limits for what is possible, that is it determines the future alternatives.
(ii) The internal control capacity of the system or connectivity. It is the degree of flexibility or rigidity between internal control variables and processes, such as their sensitivity or not to disturbance. This property determines the degree to which a system can control its own evolution.
(iii) The ability to adapt or resilience of the system. This is a measure of vulnerability to unexpected or unpredictable shocks. It is assumed that this property is opposite to the vulnerability of the system.

These properties shape the responses of ecosystems, as well as institutions and people to crises.

From the interaction of the first two properties – potential and connectivity – Holling (2001) and Holling and Gunderson (2002) established four key functions or phases of an adaptive cycle of a theoretical and complex system. They are: Exploitation (phase r), conservation (phase K), release (omega phase, Ω) and reorganization (alpha phase, α). The names of r and K phases were inspired by the concepts of r and K species, to emphasize the rapid colonization of newly disturbed areas and the slow accumulation and storage of materials and energy, respectively (Holling and Gunderson 2002).

(i) Exploitation (phase r). The potential or wealth increases incrementally together with greater efficiency, but also together with greater rigidity. The potential or wealth includes the accumulated ecological (nutrients, biomass and physical structure of the ecosystem), economic, social and cultural (skills, networks of human relations and mutual trust) capitals, as well as fortuitous mutations and inventions not expressed. In case of an ecosystem, it is observed that as it progresses to the K phase (slow sequence from exploitation to conservation), the accumulation of nutrients and biomass resources become increasingly united within the existing vegetation, preventing other competitors from using them.

(ii) Conservation (phase K). Although the system potential for other use is high, it is expropriated and controlled by specific biota and ecosystem processes in place. This results in an increase of the system connectivity and stability. As capital accumulates, the system becomes more connected, but also, more rigid in its control. As the potential increases, slow changes gradually expose the system to increasing vulnerability, which is known as a decrease in resistance to threats. Therefore, the ecosystem is approaching a situation where an accident can occur. After the system is broken, its trajectory moves abruptly backwards, from K to Ω.

(iii) Release (phase Ω). A break in the system can trigger the release of accumulated potential, also known as "creative destruction". The change is caused by threats or disturbing agents (e.g., wind, fire, diseases, insect outbreaks, drought, competitors, antagonistic groups). In the case of an ecosystem, resources that were accumulated, and sequestered, in vegetation and soil are suddenly released and close organization is lost. The Ω to α phase is a period of rapid reorganization during which new recombinations can unexpectedly give rise to situations that lead to innovations in the next cycle.

(iv) Reorganization (phase α). The low connectivity of the system allows unexpected combinations of isolated or restricted innovations in advance, that can generate new opportunities. When uncertainty is great, potential is high and controls are weak, pulses or waves of innovation are produced which, by recombination, give rise to new innovations. Some innovations fail to be tested, but others survive and adapt to r to K phase of subsequent growth. In other words, the system is renewed, or the potential can escape and change towards a less productive and organized system.

Initially, the adaptive cycle path from r to K ("frontal loop") becomes progressively more predictable as it develops. On the contrary, the phase from Ω to α ("return cycle") is inherently unpredictable and highly uncertain; at this stage previously accumulated mutations, inventions, external invaders and capital can be reorganized into new combinations, some of which generate new opportunities. Holling (2001) described it as if two separate objectives were working, but in sequence: while the first maximizes production and accumulation, the second maximizes invention and rearrangement. Since both objectives only occur sequentially, they cannot be maximized simultaneously. However, the success in achieving one of these objectives inexorably sets the stage for the other –its opposite– to be unleashed. In summary, the adaptive cycle is the interaction of two opposite processes: on the one hand, growth and stability; on the other, change and variety.

The adaptive cycle perspective has been used to analyze various systems and illustrate how human societies face changes and build adaptive capacity (Holling and Gunderson 2002; Berkes et al. 2003). In the case of pest management, Hoy (2009) described the management of pest resistance to pesticides in agroecosystems through the adaptive renewal cycle. What is interesting about Hoy's work is that it analyzed the context of pest management where pesticides play a central role, a situation that, as we have mentioned in the criticisms to IPM, prevails when IPM is implemented.

According to Hoy (2009), the growth and stability of the agroecosystem (phase from r to K or frontal loop) occurs because new pesticides are available or because new agricultural products are produced (phase r); this condition favors the pest control with acceptable gains for producer, however, the agroecosystem becomes very dependent on pesticides (phase K). As this dependence increases, the agroecosystem gradually exposes itself to a growing vulnerability (i.e. increased pest resistance to pesticides); by reducing the resistance of agroecosystem towards pests, it approaches a situation where, according to Holling (2001) and Holling and Gunderson (2002), it would cease to be profitable. For frontal loop to remain, it is assumed that pest resistance to pesticides will occur, but with availability of new chemical pesticides, it is anticipated that resistance would be manageable. Otherwise, that is, if resistance is not manageable, the return cycle (change and variety or phase from Ω to α) occurs because the pests are uncontrollable; it is then that the agroecosystem is in crisis, it becomes disorganized and financial losses occur.

Years before the proposal of adaptive renewal cycle, Smith (1971, cited by Andrews 1989) characterized the following seven historical phases of plant protection (Table 2.1):

(i) Subsistence
(ii) Incremental change
(iii) Exploitation
(iv) Crisis
(v) Disaster
(vi) Supervised multi-tactical combat
(vii) Targeted management

These phases – which largely coincide with the analysis of Hoy (2009) – highlight the close relationship of agricultural production, the use of pesticides and their consequences when used excessively, as it often happens under the IPM paradigm. Unlike Hoy (2009), Smith (1971) stated that, in response to the crisis and crop disaster caused by the high dependence on pesticides, "supervised multi-tactical combat" and "targeted management" will be used (Table 2.1) which, according to Luckmann and Metcalf (1975), is nothing other than the IPM paradigm.

Figure 2.4 integrates the historical phases of Smith's plant protection with the analysis of pesticide resistance management of Hoy (2009), according to Holling's adaptive renewal cycle. The question is whether the IPM reactive strategy – avoiding crisis and disaster caused by high dependence on pesticides – is the best strategy for pest management.

Table 2.1 The historical phases of plant protection

Phases[a]		Characteristics[a]	Adaptive renewal cycle[b]
1	Subsistence	Small producers; no modern inputs are used; low yields; local market; manual destruction of pests; use of resistant local varieties; traditional cultural practices; pest control mainly due to natural biological control; polycultures are common.	Reorganization (phase α)
2	Incremental change	Most subsistence phase procedures; occasionally pesticides are applied, or a local variety is replaced by an improved one; some portion of the product can be sold in the national market; yields can increase and, to a certain extent, they are less fluctuating.	Exploitation (phase r)
3	Exploitation	New technology packages and modern inputs are introduced on a large scale; improved varieties (more productive, but more susceptible to pests) are the central elements of changes; a lot is invested in fertilizers and pesticides (they are easy to get, cheap and sellers / extension agents promote their use); yields increase greatly; the product is sold in national and international market; there are many agronomic and social changes; the size of the farm is increased; continuous irrigation and sowing of monoculture; there is an optimistic view of the future.	Conservation (phase K)
4	Crisis	Excessive use of pesticides and abandonment of other practices, such as polyculture and crop rotation, produce resistant pests and the elimination of natural enemies results in rapid pest outbreaks, even in untreated crops; the economic and environmental costs derived from frequent use of pesticides are dramatically increased; environmental degradation occurs on farm and neighboring farms; products can be rejected in the international market due to the presence of toxic residues; acute poisonings may occur; the economic well-being of the farmer is endangered, and uncertainty prevails.	Release (phase Ω)
5	Disaster	It is no longer possible to continue crop production with current techniques due to the severity of economic and social problems.	The system is not renewed and its potential escapes and changes towards a less productive and organized system.
6	Supervised multi-tactical combat	In response to the crisis and crop disaster, pest management procedures are restructured; cultural and biological controls are used on a regular basis; insecticides are applied only if their use is economically justified; participation of pest control technicians and technical advisors is common; the procedures used are empirical and there is no deep ecological understanding of the system.	Conservation (phase K)

(continued)

Table 2.1 (continued)

Phases[a]		Characteristics[a]	Adaptive renewal cycle[b]
7	Targeted management	Pest management procedures are based on a detailed theoretical understanding of the system; decision making is facilitated through sophisticated technologies such as computer use and monitoring systems that allow high levels of prediction and planning to be achieved.	Conservation (phase K)

[a]Smith (1971, cited by Andrews 1989), Luckmann y Metcalf (1975)
[b]Holling (2001)

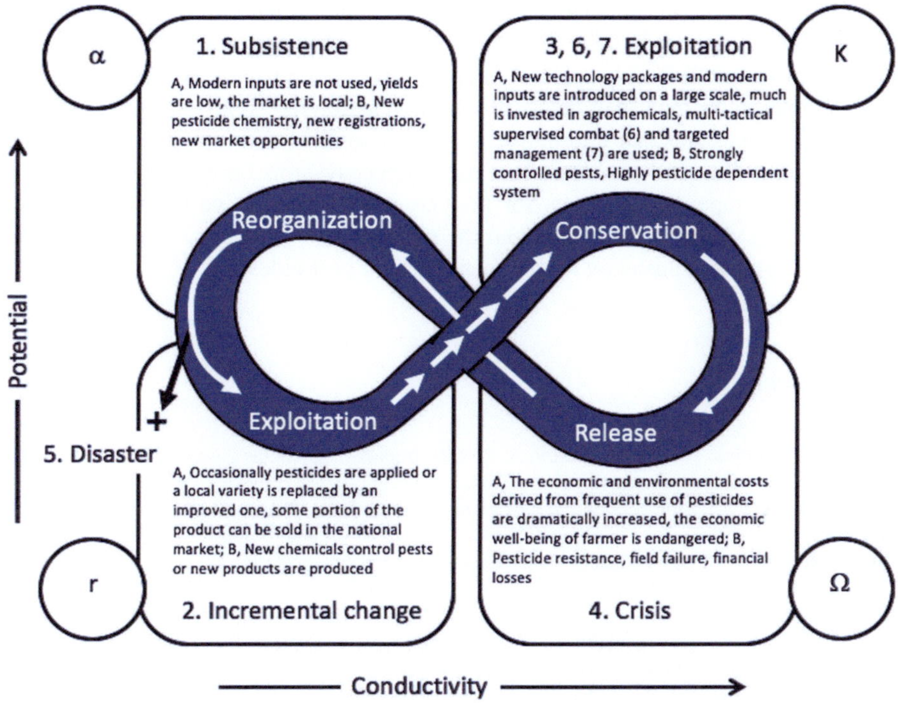

Fig. 2.4 Adaptive renewal cycle of pest management. Numbers (1–7) represent the historical phases of Smith's plant protection: 1 = subsistence, 2 = incremental change, 3 = exploitation, 4 = crises, 5 = disasters, 6 = multi-tactical supervised combat, and 7 = targeted management. A = the characteristics identified by Smith (1971); B = the characteristics identified Hoy (2009). The phases of adaptive cycle, according to Holing (2001), are represented by: r (exploitation), K (conservation), Ω (release) and α (reorganization). It avoids moving from phase K to phase Ω through supervised multi-tactical combat (6) and targeted management (7); in case these are not implemented or fail, the system quickly goes from Ω to α. The system disaster (5) arises if innovations to face failures are not available. Modified from Holling and Gunderson (2002). © 2002 From Panarchy by Lance Gunderson and C.S. Holling. Reproduced by permission of Island Press, Washington, DC

2.8 Conclusion

Emerged in response to excessive pesticides use, IPM has been in continuous evolution since Stern et al. (1959) proposed it to integrate chemical with biological control. Since then, IPM has gone from "control" concept – term that Pedigo and Higley (1996) describe as "having power over something" (i.e., chemical control) – to "management" concept which refers to the prudent use of media to achieve a desired goal. In this future, IPM gave rise to almost 70 definitions, many of them emphasizing economic and environmental aspects over social ones (Bajwa and Kogan 1996, 2002). In particular, IPM has had an old-fashioned relationship with pesticides. And it could not be otherwise, because according to some of its critics (e.g. Morse and Buhler 1997; Altieri et al. 1997; Ehler and Bottrell 2000; Altieri and Nicholls 2003), its concept favors the substitution of supplies, the search for the "silver bullet" and the justification for the use of pesticides through action thresholds. Consequently, in the last 30 years various alternative strategies to IPM that emphasize systemic approaches and natural processes that regulate ecosystems have emerged, limiting the use of therapeutic measures for pest control. The adaptive renewal cycle of Holling (2001) helps us to understand the properties of agroecosystems that shape responses to crises caused by high dependence on pesticides.

References

Aktar, W., Sengupta, D., & Chowdhury, A. (2009). Impact of pesticides use in agriculture: Their benefits and hazards. *Interdisciplinary Toxicology, 2*(1), 1–12.

Altieri, M. A. (1994). *Biodiversity and pest management in agroecosystems*. New York: Hayworth Press.

Altieri, M. A. (2018). *Agroecology: The science of sustainable agriculture* (2nd ed.). Boca Raton: CRC Press.

Altieri, M. A., & Nicholls, C. I. (2003). Ecologically based pest management: A key pathway to achieving agroecosystem health. In D. J. Rappoport, W. L. Lasley, D. E. Rolston, et al. (Eds.), *Managing for healthy ecosystems* (pp. 999–1010). Boca Raton: Lewis Publishers.

Altieri, M. A., & Rosset, P. (1996). Agroecology and the conversion of lárge-scale conventional systems to sustainable management. *International Journal of Environmental Studies, 50*, 165–185.

Altieri, M. A., Rosset, P. M., & Nicholls, C. I. (1997). Biological control and agricultural modernization: Towards resolution of some contradictions. *Agriculture and Human Values, 14*(3), 303–310.

Andow, D. A. (1991). Vegetational diversity and arthropod population response. *Annual Review of Entomology, 35*, 561–568.

Andrews, K. L. (1989). Introducción a los conceptos del manejo integrado de plagas. In K. L. Andrews & J. R. Quezada (Eds.), *Manejo integrado de plagas insectiles en la agricultura: Estado actual y futuro* (pp. 3–20). El Zamorano: Escuela Agrícola Panamericana.

Anonymous. (2007). *The life and legacy of Rachel Carson*. http://www.rachelcarson.org/. Accessed 14 Jul 2019.

Antonovsky, A. (1996). The salutogenic model as a theory to guide health promotion. *Health Promotion International, 11*(1), 11–18.

Bajwa, W. I., & Kogan, M. (1996). *Compendium of IPM definitions (CID)*. http://www.ipmnet.org/ipmdefinitions/defineIII.html#90's. Accessed 14 Jul 2019.

Bajwa, W. I., & Kogan, M. (2002). *Compendium of IPM definitions (CID) – What is IPM and how is it defined in the Worldwide Literature?* IPPC Publication No. 998, Integrated Plant Protection Center (IPPC), Oregon State University, Corvallis, Oregon.

Barfield, C. S., & O'Neil, R. J. (1984). Is an ecological understanding a prerequisite for pest management? *Florida Entomologist, 67*(1), 42–49.

Benbrook, C. M., Groth, E., III, Halloran, J. M., et al. (1996). *Pest management at the crossroads*. Yonkers: Consumers Union of United States.

Berkes, F., Colding, J., & Folke, C. (2003). Introduction. In F. Berkes, J. Colding, & C. Folke (Eds.), *Navigating social-ecological systems: Building resilience for complexity and change* (pp. 1–29). New York: Cambridge University Press.

Boller, E. F., Avilla, J., Joerg, E., et al. (Eds.). (2004). Integrated production. Principles and technical guidelines, 3rd edition. *IOBC WPRS Bulletin, 27*(2), 1–49.

Bottrell, D. R. (1979). *Integrated pest management*. Council of Environmental Quality. U.S. Government Printing Office. No. 041-011-00049-1. Washington, DC.

Bottrell, D. G., & Schoenly, K. G. (2018). Integrated pest management for resource-limited farmers: Challenges for achieving ecological, social and economic sustainability. *Journal of Agriculture Science (Cambridge), 156*, 408–426.

Bradley, B. D., Christodoulou, M., Caspari, C., et al. (2002). *Integrated crop management systems in the EU*. Amended Final Report for European Commission DG Environment. Agra CEAS Consulting, 141 p.

Carson, R. L. (1980). *Primavera silenciosa*. Barcelona: Ediciones Grijalbo.

Cate, J. R., & Hinkle, M. K. (1993). *Integrated pest management: The path of a paradigm*. The National Audubon Society. Special Report.

Corbet, P. S., & Smith, R. F. (1975). Integrated Control: A realistic alternative to misuse of pesticides? In C. B. Huffaker & P. S. Messenger (Eds.), *Theory and practice of biological control* (pp. 661–682). New York: Academic Press.

Dreves, A. J. (1996). Village-level integrated pest management in developing countries. *Journal of Agriculture Entomology, 13*(3), 195–211.

Ehler, L. E., & Bottrell, D. G. (2000). *The illusion of integrated pest management*. Issues in science and technology. http://www.issues.org/16.3/index.html. Accessed 28 Jun 2019.

Frisbie, R. E., & Smith, Jr J. W. (1991). Biologically intensive integrated pest management: The future. In J. J. Menn & A. L. Steinhauer (Eds.), *Progress and perspectives for the 21st Century*, Entomology Society of America, Centennial National Symposium (pp. 151–164). Lanham, MD.

Geier, P. W. (1966). Management of insect pests. *Annual Review of Entomology, 11*, 471–490.

Graham, F., Jr. (1970). *Since silent spring*. Boston: Houghton Mifflin.

Higley, L. G., & Pedigo, L. P. (1993). Economic injury level concepts and their use in sustaining environmental quality. *Agriculture, Ecosystems & Environment, 46*(1–4), 233–243.

Higley, L. G., & Pedigo, L. P. (1996). The EIL concept. In L. G. Higley & L. P. Pedigo (Eds.), *Economic thresholds for integrated pest management* (pp. 9–21). Lincoln: University of Nebraska Press.

Holling, C. S. (2001). Understanding the complexity of economic, ecological, and social systems. *Ecosystems, 4*, 390–405.

Holling, C. S., & Gunderson, L. H. (2002). Resilience and adaptive cycles. In L. H. Gunderson & C. S. Holling (Eds.), *Panarchy: Understanding transformations in human and natural systems* (pp. 25–62). Washington: Island Press.

Hollingsworth, C. S., & Coli, W. M. (2004). Consumer response to IPM: Potential and challenges. In O. Koul, G. S. Dhaliwal, & G. W. Cuperus (Eds.), *Integrated pest management: Potential, constraints and challenges* (pp. 255–264). Wallingford: CAB International.

Hoy, C. W. (2009). Pesticide resistence management. In E. B. Radcliffe, W. D. Hutchinson, & R. E. Cancelado (Eds.), *Integrated pest management: Concepts, tactics, strategies and case studies* (pp. 192–204). Cambridge: Cambridge University Press.

Kiritani, K. (2000). Integrated biodiversity management in paddy fields: Shift of paradigm from IPM toward IBM. *Integrated Pest Management Reviews, 5*(3), 175–183.

Kogan, M. (1998). Integrated pest management: Historical perspectives and contemporary developments. *Annual Review of Entomology, 43*, 243–270.

Lewis, W. J., van Lenteren, J. C., Phatak, S. C., et al. (1997). A total system approach to sustainable pest management. *Proceedings of the National Academy of Sciences of the United States of America, 94*, 12243–12248.

Luckmann, W. H., & Metcalf, R. L. (1975). The pest-management concept. In R. L. Metcalf & W. H. Luckmann (Eds.), *Introduction to insect pest management* (pp. 3–35). New York: Wiley.

Metcalf, R. L. (1989). Insect resistance to insecticides. *Pesticide Science, 26*(4), 333–358.

Morse, S., & Buhler, W. (1997). *Integrated pest management: Ideals and realities in developing countries*. London: Lynne Rienner Publishers.

Norris, R. F., Caswell-Chen, E. P., & Kogan, M. (2003). *Concepts in integrated pest management*. Upper Saddle River: Prentice Hall.

NRC (National Research Council). (1996). *Ecologically based pest management: New solutions for a new century*. Washington, DC: National Academy Press.

Olle, T. M. (2000). *"P" is for poison: Update on pesticide use in California schools*. Californians for Pesticide Reform. California Public Interest Research Group Charitable Trust, CA, USA.

Pedigo, L. P. (1999). *Entomology and pest management* (3rd ed.). New Jersey: Prentice Hall.

Pedigo, L. P., & Higley, L. G. (1996). Introduction to pest management and thresholds. In L. G. Higley & L. P. Pedigo (Eds.), *Economic thresholds for integrated pest management* (pp. 3–8). Lincoln: University of Nebraska Press.

Popp, J., Pető, K., & Nagy, J. (2013). Pesticide productivity and food security. A review. *Agronomy for Sustainable Development, 33*(1), 243–255.

Smith, R. F. (1971). Fases en el desarrollo del control integrado. *Boletín de la Sociedad Entomológica del Perú, 6*, 54–56.

Smith, R. F., & Reynolds, H. T. (1966). *Principles, definitions and scope of integrated pest control*. In: Proceedings of the FAO symposium. Integrated pest control. FAO-UN, Roma, Italy, pp. 11–17.

Smith, R. L., & Smith, T. M. (2001). *Ecología* (4ta. Ed). Madrid: Addison Wesley.

Stern, V. M., Smith, R. F., van den Bosch, R., et al. (1959). The integrated control concept. *Hilgardia, 29*(2), 81–101.

Tshernyshev, W. B. (1995). Ecological pest management (EPM): General approaches. *Journal of Applied Entomology, 119*, 379–381.

van den Bosch, R., Leigh, T. F., Falcon, L. A., et al. (1971). The developing program of integrated control of cotton pests in California. In C. B. Huffaker (Ed.), *Biological control* (pp. 377–394). New York: Plenum Press.

Vega, D., Gazzano Santos, M. I., Salas-Zapata, W., et al. (2019). Revising the concept of crop health from an agroecological perspective. *Agroecology and Sustainable Food Systems*. Published online: 24 Jul 2019.

Warnert, J. E. (2009). The 50th anniversary of a great idea: Landmark Hilgardia article on 'integrated control' considered 'most important' pest control paper of 20th century. *California Agriculture, 63*, 160–161.

Chapter 3
The Need for a Holistic Approach to Pest Management

There is a lack of adaptation that is ever greater, deeper and more serious among our discordant, chopped, trapped in disciplines, wisdoms and on the other hand, some increasingly multidisciplinary, transverse, multidimensional, transnational, global and planetary realities and problems.
Edgar Morin, The well-ordered mind (Translated from Morin E (2000) La mente bien ordenada: Repensar la reforma, reformar el pensamiento. Ed. Seix Barral. Barcelona, p. 13.)

Abstract This chapter includes definitions of concepts such as reductionism, holism, complexity and socio-environmental systems, among others, as a starting point to favor incorporation of a holistic approach to pest management. It emphasizes the importance of managing socio-environmental systems by setting a goal based on the quality of life that reflects the most important values of people. It is recognized that pest problems are complex, since there are many factors that affect them. As elements of a socio-environmental system, it is proposed that the solution of problems generated by pests depends on the solution of other problems or elements of the system. In turn, their own solution contributes little to solving other problems. Next, the problem of socio-environmental systems of family farming and how the poverty of many farmers limits the management of pests is presented. Finally, there is a reflection on the importance of thinking and acting holistically as a principle to change the current pest management paradigm.

Keywords Reductionism · Holism · Complexity · Family farming · Pests

J. F. Barrera, *Beyond IPM: Introduction to the Theory of Holistic Pest Management*, Sustainability in Plant and Crop Protection 15,
https://doi.org/10.1007/978-3-030-43370-3_3

3.1 Introduction

Pest problems are complex. There are many factors that affect a pest status, the damage it causes, and the actions undertaken for its management. Among them, according to Norton (1993), the following can be mentioned: crop system and agronomic factors; pesticides and other technologies; natural enemies; weather conditions; economic forces; agricultural policies; institution capacities; knowledge and perceptions of producers. For example, if we want to understand pest problems and make appropriate recommendations, many factors must be taken into account when designing and implementing management programs.

Starting from what was expressed above, we can affirm that pest management requires understanding how complex systems such as socio-environmental systems work and how to intervene in them. A socio-environmental system[1] is a set of interrelated elements that intervene in different dimensions of space and time under processes in which human beings interact with each other and with nature. This occurs according with their capabilities, their social, environmental and economic context, as well as with the internal rules of communities and those governing their links with external actors. As an example of a socio-environmental system, the coffee region of Soconusco is presented in Fig. 3.1, showing how pests may be considered as one of the many elements of the system.

Therefore, in order to carry out an effective and sustainable management of pests in socio-environmental systems it is necessary to follow a holistic approach as proposed by Barrera (2009). In other words, it is considered essential to manage pest by taking into account the socio-environmental system as a whole. Based on our experience, we can say that the lack of a holistic approach in many government programs has hindered the management of pests. To a large extent, we believe that this is due to the fact that formation of disciplinary human resources does not provide a theoretical framework and appropriate methodology to implement holistic pest management programs. In this chapter, we will focus on providing the theoretical elements that support the need to incorporate such an approach in pest management.

3.2 Concepts and Implications of the Holistic Approach

According to Hart (1985), holistic terminology entered the field of ecology with the following four concepts: the "*whole*" by Smuts in 1926; the "*holocoen*" by Friederichs in 1930; the "*biosystem*" by Thieneman in 1939; and the "*ecosystem*" by Transley in 1935.

[1] In collaboration with Lorena Soto and Balente Herrera we have developed the definition of "socio-environmental system" taking into account the definition of "global system" of García (1986) and the definition of "spaces of socio-environmental innovation" of Naranjo-Piñera et al. (2012).

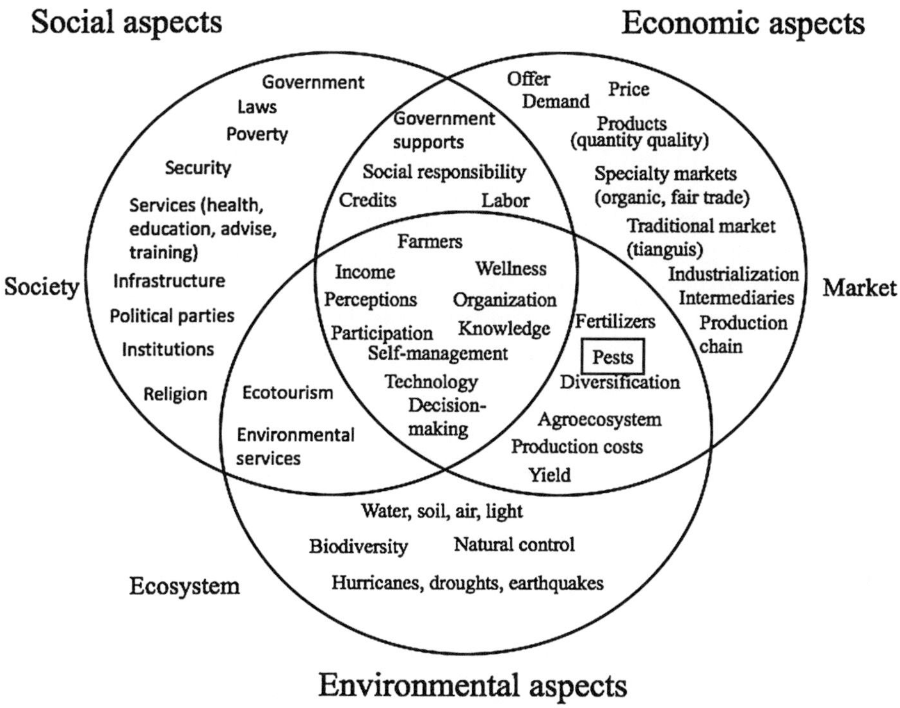

Fig. 3.1 Socio-environmental system of Soconusco coffee region in Chiapas, Mexico. The interactions between the elements of the system are shown by grouping them by social (society), economic (market) and environmental (ecosystem) aspects. Note the articulating role of the human being in the center of the figure; likewise, it can be seen that pests are one of the many elements of the system. (Modified from Barrera 2009)

The holistic approach is not new in the literature related to pest management (for example, Altieri et al. 1983; Prokopy 1987; Cate and Hinkle 1993; Benbrook et al. 1996; James et al. 2003), however, the term "holistic" is usually used in a general and abstract manner.

Two philosophies, almost always antagonistic, are important at this regard: reductionism and holism. However, rather than opposing each other, these are complementary philosophies under the holistic approach that we propose, as Mata and Pérez (2000) pointed out. In the case of disciplinary and transdisciplinary research, the latter draws on the first, which in turn is clarified in a new and fruitful way.

Next, the characteristics of both philosophies will be presented and elements on complexity will be given to better address the holistic approach and the implications it has in scientific thinking.

3.2.1 Reductionism

"Reductionism" or "atomism" is seen as the philosophy opposed to holism, since it assumes that complex systems can be explained by reducing them to their fundamental parts (Brom 1996). In ecology, the method that studies the main parts of the system first, and then tries to integrate the whole from them, is called "merological" method (from the Greek word *meros* which refers to part). According to Eugene P. Odum (1913–2002), the reductionistic approach has dominated science since Isaac Newton (1643–1727) (Odum 1986).

Scientific research from cellular and molecular levels, typical of the reductionist approach, has given many benefits to society. Nevertheless, Odum (1986) believed that science, considering only the single unit level, may contribute very little to the well-being and survival of humanity. In particular, if this "cellular" level does not interact and exchange information with the higher levels of organization it belongs to, no solution can be found to the complex social and environmental problems that affect us today. In this regard, Morin (2000) also pointed out that there is an increasingly large, deep and serious lack of adaptation between the knowledge embedded in disciplines and the increasingly multidisciplinarity that characterizes global realities or problems.

By becoming a dominant way of thinking, scientists attempted to study social processes under a reductionism perspective. Unfortunately, the results have been lean or wrong (Cocho 2014a, b). According to Morin (2000), this was due to the fact that hyperspecialization prevents seeing both the global – fragmented into disciplines – and the essential – because it dissolves it. The parable of the well-known tree exemplifies the problem of hyperspecialization (Fig. 3.2).

In the case of crop protection, Tschirley (1984) believed that the difficulty of developing strong interdisciplinary programs in USA, in which specialists from different disciplines would share their experience, was due to:

(i) An educational system that promotes specialization.
(ii) Promotion based on publication.
(iii) The creation of professional societies by disciplines.
(iv) The organization of universities by disciplinary departments.
(v) Administrators with a disciplinary approach.

3.2.2 Holism

In biology, Jan Christian Smuts (1870–1950) is considered to have introduced the term "holism" in 1926 (Hart 1985; Garduño Ochoa and Carvajal 1985; Savory and Butterfield 1999). The Greek root of holism, *holos*, means everything, whole, total. It refers to the idea that properties of a system (biological, chemical, social, economic, mental, linguistic, etc.) cannot be determined or explained only by the sum

Fig. 3.2 The parable of the well-known tree. Several scientists analyze their research objects. They are very focused on the details of their own observations, being only part of each other's interest. After finishing their investigation, each one will publish their results on the root, stem, branches, leaves or insects associated with the tree, but they do not realize that these parts form a tree because that holistic approach is only possible by taking a step back to discuss their observations with those of their colleagues. (Modified from Jørgensen and Müller 2000; Barrera 2009)

of the parts that compose it. On the contrary, the system as a whole is the one who determines in an important way how the parts work (Smuts 1936).

The holistic approach is considered a basic element in ecology as a starting point for the development of the ecosystem, its key concept (Garduño Ochoa and Carvajal 1985). The method that studies ecosystems by measuring inputs and outputs, to establish collective and emerging properties of the whole, and then investigates their parts according to different needs, is called "holological", a word also derived from *holos*. In 1915, Edward Asahel Birge (1851–1950) published a study on the calorific energy of lakes, considered the first investigation that applied the holological method (Odum 1986).

3.2.3 *The Systemic Conception*

From the perspective of systems analysis or "systemic approach", an area of work has emerged to apply the scientific method to problems related to complex systems (Grant et al. 2001).

The General System Theory of Ludwig von Bertalanffy (1901–1972) was one of the first to justify and address the study of "totality". 70 years ago, this theory pointed to the "analytical approach" – predominant in physics, biology, medicine, psychology, economics and other sciences – as opposed to the "organismic" conception (von Bertalanffy 1950). The "classical analytical method" approach, the product of scientific rationality and the axis of Western knowledge, implies the possibility of resolving an entity into parts. In addition, it assumes that the interaction of the parts is so negligible that it is enough to isolate and then add their effects (Ramírez 2014). Unlike the analytical approach, the "systemic approach" encompasses all elements of the system studied, as well as their interactions and interdependencies (de Rosnay 1975).

Approaches based on the notion of systems try to overcome the limitations of the "reduction method" (Cocho 2014a, b). According to Gutiérrez Sánchez (2014), the systems on which theories are built are a basic plot or structure where processes result from the interaction of elements integrated at different levels. Not long ago science was concerned with reducing the observed phenomena to the interaction between elementary parts. At present it focuses on the notions of totality and hierarchy, on organizational problems and on those that cannot be reduced to local events, on the relationships that emerge from the totality and that are not manifest in the behavior of its parts (Ramírez 2014).

3.2.4 The Complexity

According to Cocho (2014a, b), the systemic conception has generated many and diverse schools of thought to understand the world where we live, with a flourishing of publications on complex systems. García (1994) defined "complex systems" as situations that are characterized by the confluence of multiple processes whose interrelationships constitute the structure of a system that functions as an organized totality. The "complexity" of a system is not only given by the heterogeneity of its elements or subsystems – which, by the way, define its field discipline – but also by the interdefinability and mutual dependence of the functions that those elements carry out within the totality (García 1994). It is precisely this property that prevents analyzing a complex system by simply adding specific studies of each of the subsystems. So, the meaning of the word "complex" differs from being "complicated" or being "composed of heterogeneous elements."

The research that addresses the study of complex systems, according to García (1994), is "interdisciplinary research". However, partial studies of some of its elements or functions are not excluded, since not all research is interdisciplinary. In this regard, Odum (1986) is persistent in pointing out that holistic and reductionist methods should be complementary and not antagonistic, and should be equally important simultaneously, not alternately.

With the study of complex systems, the "new sciences of complexity" emerged, which seeks to establish a dialogue between natural and exact sciences with social and human disciplines (Maldonado 2005). It is intended to give rise to a general theory of knowledge that integrates physical objects and human beings in their cultural and social context (de la Peña 2001).

There are more and more works that in various fields of science aimed at building an epistemological paradigm on complexity as an alternative to the dominant paradigm based on simplicity (Munné 2004). Edgar Morin's work, "The Method" is a complete treatise on the study of complexity in six volumes, where he warns of the need to find a method that – unlike the Cartesian scientific method – detects and does not hide unions, joints, solidarity, implications, overlaps, interdependencies and complexities of reality and the human condition (Morin 2006).

3.2.5 System Limits

No system is defined *a priori*, at the starting point of an investigation. In fact the system is not defined, but is definable (García 1986). The most adequate definition of the system will arise during the course of its investigation.

Investigating a complex system means studying a "piece of reality" with all its physical, biological, social, economic and political aspects (García 1986). Based on the above, the definableness of a system where pests constitute one of its elements, can go beyond the agricultural system environment if the impact of pests determines – among others – market for products (e.g., damage to quality), people's health (e.g., contamination with pesticides), or producers' cooperation in government programs (e.g., their perceptions of pests). This means that the study of pests should not be limited to specific studies on their biology, behavior or control – actions that define much of researchers' work – but also, to establish the causal relationships between them and other system components.

In reality, complex systems lack precise limits, both physically or geographically. In terms of problems, the conceptual framework or the different phenomena with their spatial and temporal scales leads to establishing more or less arbitrary "partitions". Therefore, it is necessary to establish selective criteria with respect to interactions between system elements, since no study could cover all of them (García 1986).

To the extent that they mutually determine each other, the components of a system cease to be independent, and conform their structure (García 1986). Therefore, asking yourself *Why a pest is a pest*?, invites you to identify the elements of a system that, when interacting, determine the population outbreaks of an organism (Lewis et al. 1997). Other similar questions are: *What are the producers' perceptions of pests? What is the priority they assign to them? What knowledge do they have about its management? What are they doing to fight them? What limits them for effective management?* Although these questions are characteristic of the "pest" element, they raise the search for interactions with other key elements of the system.

In the case of coffee producers in Chiapas, it was proposed to define their system with the question: *What are the main problems that limit producing coffee in a sus-*

tainable way? (Barrera 2006). A question of this nature allows to cover the three levels of processes identified by García (1986) to analyze the dynamics of the systems, namely: first level or basic processes (e.g., losses caused by pests); second level processes or metaprocesses (e.g., biological control of pests); and, third level processes (e.g., access policies to organic coffee market).

3.2.6 The Holistic Goal

Once the limits have been established, the management of the system acquires a meaning by establishing a goal based on the defined whole. This goal, called "holistic goal" by Savory and Butterfield (1999), is a declaration of principles that establishes what is to be achieved in the short and long term. According to the aforementioned authors, the holistic goal must express the values of people, that is, what motivates them to perform certain actions. Only a goal based on the quality of life that reflects the most important values, can generate in people the degree of commitment necessary to achieve what they want.

The holistic goal has three fundamental components: the quality of life or how people want to live to achieve together, based on their values, whatever they set; the forms of production to produce what will allow them to reach that quality of life and prosper in their economic activities; and the future resource base for sustainable production (Savory and Butterfield 1999).

3.3 Interaction Between System Elements

The analysis of the 10 most important problems arising from the question "*What are the main problems that limit producing coffee in a sustainable way in Chiapas?*" relies on "structural analysis",[2] an analytical technique determining the interaction among problems (elements) of a system (Mojica 1991, 2004). It resulted in pest management strongly dependent on solving other problems such as organization of producers and training. Already at an early time – in the 1970s – it was recognized that pest control would depend largely on external factors proceeding from other sectors or disciplines (Corbet and Smith 1975). Moreover, the structural analysis indicated that the solution to a problem caused by pests contributed little to solve other system problems (Barrera et al. 2004).

Through workshops held in several locations where we applied the structural analysis technique (Barrera et al. 2013), we find that the problems identified by the participants were not always the same. However, in those locations that identified pests as one of the 10 main problems, the pests invariably were placed in the "exit

[2] Chapter 5 presents the methodology of structural analysis.

zone" of this analysis (Table 3.1). These results showed a clear need to carry out pest management under a holistic approach.

3.4 Socio-environmental Systems of Family Farm

Today, the Food and Agriculture Organization of the United Nations (FAO) and the International Fund for Agricultural Development (IFAD) recognize that the world food system not only failed to eliminate hunger and poverty, but also generated deep social inequalities (FAO and IFAD 2019). This problem increases impacts of climate change and degradation of natural resources (water scarcity, soil depletion, loss of biodiversity), leading to innumerable conflicting issues such as the increase in pests.

It is estimated that livestock producers, fishermen, foresters, indigenous populations and other groups of food producers who practice family farm[3] in developed and developing countries provide more than 80% of the world's food[4] (FAO 2014); in addition, family farmers are the main investors in agriculture and constitute the backbone of the rural economic structure (FAO and IFAD 2019).

However, family farmers often have limited access to natural resources, production assets and markets, which places them among the poorest and most vulnerable groups in society at economic, financial, social and environmental risks, particularly women, youth, indigenous peoples and ethnic minorities (FAO and IFAD 2019). It is estimated that 70% of the poorest live in the countryside and their well-being depends on agriculture (IFAD 2013).

In Mexico, Regalado López et al. (1996) found that the use of technology by corn producers in the state of Puebla was related to personal factors such as age, schooling, family size and sex. Another finding was that the producers who had less land surface, who did not have their own machinery, who did not receive technical assistance, and who did not participate in a producer organization, did not use corn productivity technology more accurately. In other words, as Meerman et al. (1996) showed, in order to overcome the bottlenecks of agricultural development so that the farmer has a capacity to invest, the problems of socioeconomic origin must first be reduced.

[3] The concept of family farming, according to FAO and IFAD (2019), is a way of organizing agricultural, forestry, fisheries, livestock and aquaculture production that is managed by a family and depends mainly on the capital and labor of its members, both women and men. The family and the farm are related to each other, evolve together and combine economic, environmental, social and cultural functions; in short, family farm is agriculture that includes all family-centered farming activities.

[4] A more recent study estimates that farms with less than 2 ha produce globally between 28 and 31% of total crop production and between 30 and 34% of food supply in 24% of the gross agricultural area (Ricciardi et al. 2018).

Table 3.1 Problems identified in several coffee production locations in Mexico and their distribution according to the structural analysis zone. The numbers correspond to the locations where the workshops were held[a]. The problems were sorted by type[b]

Problems	Type of problem	Structural analysis zone[c] Power	Conflict	Exit	Autonomous problems
Lack of activity planning	Ad		4		
Bad administration of government supports to farmers	Ad		7		
Soil erosion	Am			2, 5	
Weather alteration	Am				5
Unfavorable climatic factors	Am				7
Unfavorable markets	Co	2, 6	1, 5		
The purchase and sale contracts are not fulfilled	Co			8	
Dishonest local coffee shoppers	Co	5			
There is no culture about coffee quality	Cu			1	
Government paternalism	Cu				7, 8
Mistrust in investing in coffee	Ec		3		
Instability and low coffee prices	Ec	2, 3, 8	1, 5		6
Insufficient economic resources	Ec	2	1, 3, 4, 7, 8	6	
Government support out of time	Ec	4, 5			1, 6
Poor training	Ed	1, 4	2, 6	3	
Resistance to using new technologies	Ed		7		
Fruit trees unproductive due to pesticides	Ed				2
Family disintegration by migration	F			2	
Family problems due to inheritance	F	7			
Lack of infrastructure for coffee processing	I			7	4, 8
Plantations located far from communities	I			3	
Producer disorganization	O	1, 3, 4, 6	5		
Disarticulated industrialization	O	6			1
Poor technical assistance	PS	1, 4, 8	2, 5	6	
Lack of labor	PS				1, 3, 4
Producer health	S	3			
Alcoholism	S	7, 8			
Lack of knowledge about coffee cultivation	T			7	

(continued)

Table 3.1 (continued)

Problems	Type of problem	Structural analysis zone[c] Power	Conflict	Exit	Autonomous problems
Lack of coffee quality control	T			1	
Pests and diseases	T			1, 2, 3, 4, 5, 6	
Old coffee plantations	T			5	
Poorly grown coffee plantations	T			6	
Abandoned coffee plantations	T			8	
Lack of organic fertilizers	T			2	
Lack of responsibility to soil management	V	2			
Lack of interest to growth coffee	V		3, 4, 7, 8		
The producer needs to value what he what he owns	V		8		

Elaboration based on data from Barrera et al. (2013)

[a]Location where the workshop was held: *1* Agustín de Iturbide, Chiapas, *2* Palestina, Chiapas, *3* Los Ángeles, Chiapas, *4* Ángel Albino Corzo, Chiapas, *5* Xalapa, Veracruz, *6* survey in Chiapas, *7* Chicomuselo, Chiapas, *8* Ocozocoautla, Chiapas

[b]Type of problem: *O* organizational, *Ad* administrative, *T* technological, *Cu* cultural, *Ec* economic, *Am* environmental, *I* infrastructure, *S* health, *Ed* educational, *PS* service providers, *V* values, *F* family, *Co* commercial

[c]It refers to the four areas of structural analysis. Power zone: Problems with high mobility and low interdependence; conflict zone: Problems with high mobility and high interdependence; autonomous zone: Problems with low mobility and low interdependence; exit zone: Problems with low mobility and high interdependence

Poverty is, in short, the condition that most limits the management of pests in the socio-environmental systems of family farm. Therefore, the fight against pests must begin with actions aimed at reducing poverty. Although it is understood that to get out of the vicious cycle of poverty means more than "having little savings" (FAO and IFAD 2019), improving income becomes a fundamental action of a virtuous circle to access sustainable livelihoods and reduce vulnerability of households of family farmers (Fig. 3.3).

3.5 Think and Act Holistically

Human beings learn to think and act holistically in a natural way. From childhood, we receive and process through our senses all kinds of information from the environment that surrounds us, and then make decisions that meet our needs. Awake or asleep we perceive all kinds of stimuli that shape our knowledge, behavior and character. Our direct or indirect contact with that environment, near or far, shallow or deep, builds our reality, and the answers to that reality make us the beings that we are and provide us with the intellectual and material resources we have to face life.

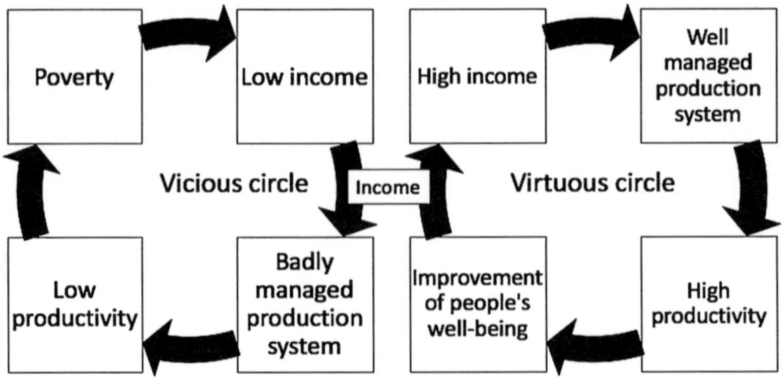

Fig. 3.3 Diagram showing that under a strategy to improve the income of producers it is possible to transform the vicious circle of poverty into a virtuous circle of well-being. It is assumed that the income improvement would increase the resources to grow their plantations, which, in turn, would increase productivity and profits. (Modified by Barrera et al. 2007)

According to Collins et al. (1989), skills and knowledge are learned in the social and functional context. Obviously, we could not survive alone for a long time if the decisions to be made to solve our problems and needs were based on reduced or incomplete information from that environment.

Paradoxically, the trajectory of higher education systems of recent times has been grounded in disciplinary logic (Morin 2000; Lanz and Fergusson 2005). During our school and university education, reality is presented to us in a fractional and disintegrated way: we usually receive information by subjects or disciplines that almost never meet again to interact, nor connect to be used in real situations (Brown et al. 1989; Collins et al. 1989; Capra 1998; Morin 2000; Luzzi 2003). This unnatural formation not only deforms our thinking and our actions in life, but also makes us unable to contribute successfully to solution of complex problems that concern society.

Possibly the most important challenge for specialists (entomologists, pathologists, etc.) of agricultural and forestry health to dare, or practice the holistic approach, is to change their attitude and approach to study or manage complex systems. And it is not for less, since the 1940s and following the trends of education and agricultural extension in the USA promoted by the Rockefeller Foundation – and which would eventually give rise to the Green Revolution – universities and research centers in Latin America have privileged the creation of departments by specialties and the training of specialists by disciplines (Jiménez-Velázquez 1990; Díaz-Sánchez 2014), which has developed and strengthened the reductionist approach. The criticism of this system is not so much the training of specialists, but the training of de-contextualized specialists of reality instead of specialists with a clear vision of the system, making it impossible for them to understand and, therefore, contribute to solution of social problems.

Because of the above and following Morin (2000), thinking and acting holistically implies the general ability to raise and treat problems; and have the organizational principles that allow us to unify knowledge and make sense of it.

Holistic thinking and action must always consider improving the well-being and capacities of farmers, since only having solved their most pressing problems, they will worry and deal with the management of pests. Thought and reverse action, which proclaims the IPM approach, will hardly allow satisfactory management of pests by farmers in the short, medium and long terms. If we want to go beyond IPM, we have to dare to think and act holistically.

3.6 Conclusion

Pest problems are complex, that is, they have many interactions with other elements of the socio-environmental system. Such interaction is peculiar because the solution of a pest problem depends heavily on solving other system problems. Moreover, its own solution contributes very little to other problems overcoming its limitations. Therefore, a pest problem management strategy requires addressing the solution of those problems that prevent an effective and sustainable management. One of the most important problems limiting pest management is the insufficiency of financial resources. Therefore, an important part of the pest management strategy should consider improving the income of producers. This reality raises the need for pest management with a holistic approach where the human being, as the articulating axis of the socio-environmental system, plays a central role.

References

Altieri, M. A., Martin, P. B., & Lewis, W. J. (1983). A quest for ecologically based pest management systems. *Environmental Management, 7*, 91–100.

Barrera, J. F. (2006). Manejo holístico de plagas: Hacia un nuevo paradigma de la protección fitosanitaria. In J. Pohlan, L. Soto, & J. Barrera (Eds.), *El cafetal del futuro: Realidades y Visiones* (pp. 63–82). Aachen: Shaker Verlag.

Barrera, J. F. (2009). La necesidad del enfoque holístico en el manejo de plagas. In J. C. Arrivillaga, M. El Souki, & B. Herrera (Eds.), *Enfoques y temáticas en entomología* (XXI Congreso Venezolano de Entomología) (pp. 1–28). Caracas: Sociedad Venezolana de Entomología.

Barrera, J. F., Parra Vázquez, M., Herrera Hernández, O. B., et al. (2004). *Plan Estatal de Manejo Agroecológico del Café en Chiapas: Guía hacia una cafeticultura sustentable*. Tapachula: Comisión para el Desarrollo y Fomento del Café de Chiapas and El Colegio de la Frontera Sur, 164 p.

Barrera, J. F., Herrera, J., & Gómez, J. (2007). Riesgo-vulnerabilidad hacia la broca del café bajo un enfoque de manejo holístico. In J. F. Barrera, A. García, V. Domínguez, & C. Luna (Eds.), *La Broca del Café en América Tropical: Hallazgos y Enfoques* (pp. 131–141). Tapachula: Sociedad Mexicana de Entomología and El Colegio de la Frontera Sur.

Barrera, J. F., Gómez-Ruiz, J., Parra-Vázquez, M. R., et al. (2013). Análisis de problemas ocasionados por plagas del café bajo un enfoque holístico. *Entomol Mex, 12*(2), 1128–1133.

Benbrook, C. M., Groth, E., III, Halloran, J. M., et al. (1996). *Pest management at the crossroads*. Yonkers: Consumers Union of United States, Inc..

Brom, Y. (1996). *Dos modos de aprender el mundo*. Perfiles educativos 74. Centro de Estudios sobre la Universidad. Universidad Nacional Autónoma de México. https://www.redalyc.org/pdf/132/13207407.pdf. Accessed 24 Oct 2019.

Brown, J. S., Collins, A., & Duguid, P. (1989). Situated cognition and the culture of learning. *Educational Research, 18*, 32–42.

Capra, F. (1998). *La trama de la vida, una nueva perspectiva de los sistemas vivos*. Barcelona: Ed. Anagrama SA.

Cate, J. R., & Hinkle, M. K. (1993). *Integrated pest management: The path of a paradigm*. Washington, DC: The National Audubon Society. Special Report. December 31, 1993, 43 p.

Cocho, G. (2014a). Presentación. In S. Ramírez (Coord), *Perspectivas en las teorías de sistemas* (2nd ed., pp. 13–17). Mexico: Universidad Nacional Autónoma de México. Biblioteca Aprender a Aprender.

Cocho, G. (2014b). Sobre la contribución de Progogine, Haken, Atlan y el Instituto de Santa Fe al estudio de la dinámica de los sistemas complejos. In S. Ramírez (Coord), *Perspectivas en las teorías de sistemas* (2nd ed., pp. 57–62). Mexico: Universidad Nacional Autónoma de México. Biblioteca Aprender a Aprender.

Collins, A., Brown, J. S., & Newman, S. E. (1989). Cognitive apprenticeship: Teaching the crafts of reading, writing, and mathematics. In L. B. Resnick (Ed.), *Knowing, learning, and instruction: Essays in honor of R. Glaser* (pp. 453–493). Hillsdale: Erlbaum.

Corbet, P. S., & Smith, R. F. (1975). Integrated control: A realistic alternative to misuse of pesticides? In C. B. Huffaker & P. S. Messenger (Eds.), *Theory and practice of biological control* (pp. 661–682). New York: Academic.

de la Peña, J. A. (2001). La complejidad de la complejidad. *Cinta de Moebio, 10*, 2–13.

de Rosnay, J. (1975). *Le macroscope, vers une vision globale* (p. 314). Paris: Ed. Seuil.

Díaz-Sánchez, S. (2014). Agrarismo, Industrialismo. Rumbo a la era de la información. In L. Victorino-Ramírez & S. Díaz-Sánchez (Coord), *Educación agrícola superior: Cambio de época* (pp. 77–150). Mexico: Centro de Estudios para el Desarrollo Rural Sustentable y la Soberanía Alimentaria. Honorable Cámara de Diputados. LXII Legislatura. Congreso de la Unión.

FAO (Organización de las Naciones Unidas para la Alimentación y la Agricultura). (2014). *El estado mundial de la agricultura y la alimentación*. Roma: La innovación en la agricultura familiar.

FAO and IFAD (Organización de las Naciones Unidas para la Alimentación y la Agricultura, and Fondo Internacional de Desarrollo Agrícola). (2019). *Decenio de las Naciones Unidas para la agricultura familiar 2019–2028*. Roma: Plan de acción mundial.

García, R. (1986). Conceptos básicos para el estudio de sistemas complejos. In E. Leff (Coord), *Los problemas del conocimiento y la perspectiva ambiental del desarrollo* (pp. 45–71). Mexico: Siglo XXI Editores, S.A. de C.V.

García, R. (1994). Interdisciplinariedad y sistemas complejos. In E. Leff, R. García, P. Gutman, V. M. Toledo, H. M. C. Vessuri, R. Fernández, & R. Brañes (Eds.), *Ciencias Sociales y Formación Ambiental* (pp. 85–124). Mexico: Gedisa Editorial.

Garduño Ochoa, R., & Carbajal, R. (1985). Hacia un enfoque de sistemas biológicos. Vol. 1. In R. Garduño Ochoa (Coord), *Biología y pensamiento de sistemas: Una aproximación bibliográfica* (p. 121). Mexico: Consejo Nacional de Ciencia y Tecnología.

Grant, W. E., Marín, S. L., & Pedersen, E. K. (2001). *Ecología y manejo de recursos naturales: Análisis de sistemas y simulación*. San José: Ed. Agroamérica. Instituto Interamericano de Cooperación para la Agricultura, 340 p.

Gutiérrez Sánchez, J. L. (2014). Teorías, sistemas y comprensión del mundo. In S. Ramírez (Coord), *Perspectivas en las teorías de sistemas* (2nd ed., pp. 107–114). Mexico: Universidad Nacional Autónoma de México. Biblioteca Aprender a Aprender.

Hart, R. D. (1985). *Conceptos básicos sobre agroecosistemas*. Turrialba: Centro Agronómico Tropical de Investigación y Enseñanza, 159 p.

IFAD (Fondo Internacional de Desarrollo Agrícola). (2013). *Smallholders, food security, and the environment*. Roma: FIDA.

James, B., Neuenschwander, P., Markham, R., et al. (2003). Bridging the gap with the CCIAR systemwide program on Integrated Pest Management. In K. M. Maredia, D. Dakouo, & D. Mota-Sanchez (Eds.), *Integrated Pest Management in the global arena* (pp. 419–434). Wallingford: CAB International.

Jiménez-Velázquez, M. A. (1990). La Fundación Rockefeller y la investigación agrícola en América Latina. *Comercio Exterior, 40*(10), 968–975.

Jørgensen, S. E., & Müller, F. (2000). Towards a unifying theory. In S. E. JØrgensen & F. Müller (Eds.), *Handbook of ecosystem theories and management* (pp. 465–471). Boca Raton: Lewis Publishers.

Lanz, R., & Fergusson, A. (2005). *La reforma universitaria en el contexto de la mundialización del conocimiento (Documento Rector)*. Paris: Observatorio Universitario de Reformas Universitarias, ORUS-Int.

Lewis, W. J., van Lenteren, J. C., Phatak, S. C., et al. (1997). A total system approach to sustainable pest management. *Proceedings of the National Academy of Sciences of the United States of America, 94*, 12243–12248.

Luzzi, D. (2003). La "ambientalización" de la educación formal. Un diálogo abierto en la complejidad del campo educativo. In E. Leff (Coord), *La complejidad ambiental* (2da ed., pp. 158–192). Mexico: Siglo XXI Editores, SA de CV.

Maldonado, C. E. (2005). ¿En qué sentido puede hablarse de diálogo de las ciencias? Acerca de las nuevas ciencias de la complejidad. *Revista de la Academia Colombiana de Ciencias Exactas, Físicas y Naturales, 29*, 417–428.

Mata, B., & Pérez, J. R. (2000). Interdisciplina y otros conceptos en juego. In B. Mata (Coord), *Retos de la investigación interdisciplinaria en el medio rural. Memoria del Primer Seminario del Centro Interdisciplinario de Investigación y Servicio para el Medio Rural* (pp. 15–20). Chapingo: Universidad Autónoma Chapingo.

Meerman, F., Van De Ven, G. W. J., Van Keulen, H., et al. (1996). Integrated crop management: An approach to sustainable agricultural development. *International Journal of Pest Management, 42*(1), 13–24.

Mojica, S. F. (1991). *La prospectiva; técnicas para visualizar el futuro*. Bogotá: Fondo Editorial Legis.

Mojica, F. J. (2004). *El modelo prospectivo llevado a la práctica. Primera versión. Convenio Andrés Bello (CAB)*. Serie: Documentos de Ciencia, Tecnología e Innovación de los países del CAB. Bogotá, 178 p.

Morin, E. (2000). *La mente bien ordenada: Pensar la reforma, reformar el pensamiento*. Barcelona: Ed. Siex Barral S.A, 185 p.

Morin, E. (2006). *El Método* (7a ed., p. 448). Madrid: Ediciones Cátedra.

Munné, F. (2004). El retorno de la complejidad y la nueva imagen del ser humano: Hacia una psicología compleja. *Revista Interamericana de Psicología, 38*, 23–31.

Naranjo-Piñera, E. J., Álvarez, D., Estrada, E., et al. (2012). ¿Qué son los espacios de innovación socioambiental? In E. Bello-Baltazar, E. J. Naranjo, & R. Vandame (Eds.), *La otra innovación para el ambiente y la sociedad en la frontera sur de México* (pp. 25–28). Mexico: El Colegio de la Fronrera Sur.

Norton, G. A. (1993). Philosophy, concepts and techniques. In G. A. Norton & J. D. Mumford (Eds.), *Decision tools for pest management* (pp. 1–21). Wallingford: CAB International.

Odum, E. P. (1986). *Fundamentos de ecología*. Mexico: Nueva Editorial Interamericana, S.A. de C.V, 422 p.

Prokopy, R. J. (1987). Holistic Pest Management: Ecological theory and Integrated Pest Management practice. Book reviews. *Science, 238*, 410–411.

Ramírez, S. (2014). Teoría general de sistemas de Ludwig von Bertalanffy. In S. Ramírez (Coord), *Perspectivas en las teorías de sistemas* (2nd ed., pp. 23–36). Mexico: Universidad Nacional Autónoma de México. Biblioteca Aprender a Aprender.

Regalado López, J., Jiménez Sánchez, L., Laird, R. J., et al. (1996). Factores asociados a la utilización de la tecnología de alta productividad entre productores de maíz de temporal. *Agrociencia, 30*(1), 139–145.

Ricciardi, V., Ramankutty, N., Mehrabi, Z., et al. (2018). How much of the world's food do smallholders produce? *Global Food Security, 17*, 64–72.

Savory, A., & Butterfield, J. (1999). *Holistic management, a new framework for decision making* (2nd ed.). Washington, DC: Island Press, 616 p.

Smuts, J. C. (1936). *Holism and evolution* (3rd ed.). London: McMillan and Co., Ltd., 358 p.

Tschirley, F. H. (1984). Integrated pest management. *Bioscience, 34*(2), 69.

Von Bertalanffy, L. (1950). An outline of general system theory. *The British Journal for the Philosophy of Science, 1*, 134–165.

Chapter 4
Theoretical Support of Holistic Pest Management

"The ultimate goals [of pest management] are not the pest, or even the crop, but the viability of rural life as a whole and a safe, sustainable food supply in the face of many kinds and increasing intensity of uncertainty. The central intellectual problem is the understanding of the behavior of complex, multilevel systems sufficient to design systems that are productive, economical, resistant to uncertainties, safe for producers and consumers, and protective of the local and global environment."
Richard Levins, From simple IPM to the management of agroecosystems (Levins R (2007) From simple IPM to the management of agroecosystems. In: Kogan M, Jepson P (eds) Perspectives in ecological theory and Integrated Pest Management. Cambridge University Press, Cambridge, UK, p. 62.)

Abstract This chapter presents the most outstanding aspects of the theory of Holistic Pest Management (HPM). The reasons why HPM emerges as a new paradigm for protection of plants, animals and human beings are explained. This paradigm draws on the convergence of three schools of thought: holistic management (system approach), agroecology (agricultural production and conservation of natural resources) and complex thinking (emphasis on interactions). It is mentioned that unlike Integrated Pest Management (IPM), which makes decisions based on pest densities, decisions in the HPM are made based on pest risk, which means putting in the center of the system the producers and their families. Thus, risk reduction involves not only preventing damage from pests, it also means reducing vulnerability and increasing the response capacity and resilience of the system. To achieve this, a process that starts thinking and acting holistically is proposed. This means that farmers promote various actions to improve their income and well-being with support of specialists who facilitate their processes, products and services. This requires a pest management program that promotes the organization, participation

J. F. Barrera, *Beyond IPM: Introduction to the Theory of Holistic Pest Management*, Sustainability in Plant and Crop Protection 15,
https://doi.org/10.1007/978-3-030-43370-3_4

and self-management of farmers, to enable them to take advantage of knowledge and technologies to prevent, rather than control, pests.

Keywords Integrated Pest Management · Holistic management · Agroecology · Complex thinking · Pest risk

4.1 Introduction

About 20 years ago we asked ourselves why it was so difficult to control the coffee berry borer (*Hypothenemus hampei*) in Mexico, a major coffee pest in the world. More specifically, we wondered why many producers did not use the available control methods (Table 4.1), although the majority (84%) claimed to suffer production losses due to this pest (Jiménez et al. 2000).

Subsequently, when comparing perceptions, knowledge and activities of coffee pest management between organic and conventional producers in Soconusco, Chiapas, we find results similar to those mentioned above (Segura et al. 2004; Jarquín-Gálvez et al. 2005, 2006). However, organic coffee producers recorded more knowledge and more pest control actions. This difference, according to Segura et al. (2004), was explained because organic coffee producers had better technical assistance and received a better price for the sale of coffee.

Table 4.1 Tactics and strategies for management of coffee berry borer (*Hypothenemus hampei*) used by coffee producers in the municipality of Tapachula, Chiapas, Mexico, 1998–1999[a]

		Producers	
Tactic	Strategy	Number	%
Cultural control[b]	Prevention	38	33.9
Chemical control[c]	Suppression	2	1.8
Cultural control + chemical control	Suppression	3	2.7
Biological control[d]	Prevention	1	0.9
Subtotal (perform some kind of control)	Prevention or suppression	44	39.3
Does not control	Coexistence	68	60.7
Total		112	100.0
Sampling of field infestation[e]	Decision making		8.0

[a]Modified from Jiménez et al. (2000)
[b]Mainly collection of infested coffee berries after harvest
[c]Use of endosulfan insecticide, but in no case the decision to apply the insecticide took into account sampling of pest infestation and action thresholds
[d]Parasitoid *Cephalonomia stephanoderis*
[e]In no case the pest infestation was quantified with the methods recommended by technical assistance institutions

When we analyze in greater depth the situation of coffee berry borer control through a diagnosis made in Chiapas (Barrera et al. 2004), we could find out that deficiencies in control of this pest were shared not only by many Mexican producers, but also by producers from other countries. Therefore, we conclude that coffee producers have a high vulnerability and little response capacity to pests such as the coffee berry borer.

Years later, in Mexico and Central America there was a strong attack of coffee leaf rust (*Hemileia vastatrix*) that destroyed many coffee plantations and brought much misery in the region (Barrera et al. 2013; Avelino et al. 2015). Among the causes that favored the intensity of the coffee disease, the following stood out: cultivation of coffee varieties susceptible to the pathogen (despite availability of resistant varieties), presence of old and poorly nourished plantations, and insufficient technical assistance and resources to cope with the disease (despite the support of Government programs). In summary, the surprise attack of the rust showed an excessively vulnerable and very limited coffee industry to respond to this kind of threats.

From these facts, we understood the urgent need to change the approach to control pests. In particular, we have found it obvious that Integrated Pest Management (IPM) has not been the most appropriate approach. This statement is even more true in the case of developing countries, where IPM is not frequently used (Morse and Buhler 1997; Morse 2009; Parsa et al. 2014).

In reviewing the literature on the subject, we found that our concern was neither new nor unique. We even found that some researchers such as Lester E. Ehler (1946–2016) and Dale G. Bottrell were very critical towards IPM (Ehler and Bottrell 2000). The agreement of these and other authors with our ideas reinforced the conviction that we had to find an alternative approach to pest management.

Such an alternative approach is Holistic Pest Management (HPH), considered a new pest management paradigm (Barrera 2006, 2020). This chapter presents the origin, concept, objective, principles, strategy and goal of HPM; likewise, the components of the virtuous circle of HPM are described and founded.

4.2 Origin

HPM has its origin in the convergence of ideas, concepts, philosophies and experiences of several schools of thought that have been concerned with the welfare of society and the conservation of the environment. The following three schools of thought have inspired us to build HPM:

- The holistic management of Dr. Allan Savory, who was one of the first to conceptualize a framework for decision-making based on the Holistic Management paradigm (Savory and Butterfield 1999).
- The agroecology that has Dr. Miguel A. Altieri as one of its most important exponents (Altieri 1987, 1994).

– The complex thinking and thought reform by Dr. Edgar Morin (Morin 2000, 2006).

HPM is strengthened with the principles of these schools of thought to build an alternative approach to reductionism that prevails in IPM.

4.2.1 Allan Savory and Holistic Management

Allan Savory is a wildlife biologist native to Zimbabwe, who developed the concept of "Holistic Management" (HM) to solve the problem of desertification. HM was defined as an approach to manage natural resources that promotes biodiversity, improves production and generates financial strength; it improves the quality of life while preserving the environment (Savory and Butterfield 1999; Savory Global 2019). Savory founded the Center for Holistic Management in 1984 and in 1999 he published his book *Holistic Management*, an updated version of the book *Holistic Resource Management* that he published in 1988 (Savory and Butterfield 1999).

Savory points out that a holistic approach is essential in management of real situations. If the decision-making were based on any other approach, he says, very different results would be obtained from those that are intended, because only the "whole" consideration leads to reality. The ideas that underlie HM only take meaning and power in relation to an objective, called by Savory "the holistic objective". As we will see later, the holistic objective enables the word "pest" (P) in HPM.

4.2.2 Miguel A. Altieri and Agroecology

Globally, agroecology has Dr. Miguel A. Altieri as one of its most important exponents.[1] According to Altieri (1987), agroecology or agricultural ecology is a theoretical framework that has the purpose of understanding agricultural processes in the broadest way; therefore, agroecology moves away from the disciplinary approach and concerns about products, to direct its attention to complex interactions between people, crops, soil and livestock.

Agroecology, Altieri (1987) tells us, emphasizes the restoration of natural pest control mechanisms, mainly through management of vegetation, which also conserves energy, improves soil fertility, minimizes risks erosion and reduces dependence on external inputs.

[1] Dr. Stephen R. Gliessman is another scientist who stands out worldwide for his teaching, research and experience in agroecological production. His textbook, with prologue of Dr. Eugene P. Odum, is an essential work for those interested in agroecology (Gliessman 2002).

The most important objective of agroecology, Altieri and Nicholls (2003) point out, is to improve biological efficiency, conserve biodiversity and maintain productivity, all while strengthening the self-sustainable capacity of an agroecosystem.

On the other hand, Bruil et al. (2019) mention that agroecology is based on application of ecological principles for the design and management of sustainable food systems where farmers and citizens are at the center of food governance. To a large extent – these authors express – agroecology is an alternative to large-scale, industrialized, specialized and technological agricultural development, which is driven from the bottom up by civil society and social movements.

Currently, agroecology has made its way between the Green Revolution and the Fourth Industrial Revolution to be among the programs promoted by FAO (Bruil et al. 2019).

4.2.3 Edgar Morin and Complex Thinking

The French philosopher, anthropologist and sociologist Edgar Morin, promoter of complexity and complex thinking, in his transcendental work "The Method" – six volumes – criticizes the Cartesian scientific method by the absolute disjunction of the object and the subject, while warning of the historical need of finding a method that detects and does not hide the unions, articulations, solidarity, implications, imbrications, interdependencies and complexities of reality and human condition (Morin 2006).

In his book *The well-ordered mind* (Morin 2000), he pointed out that hyperspecialization – the product of the disciplinary approach – prevents seeing the global (which fragments into plots) and the essential (which dissolves), when essential problems are never plots and global problems are increasingly essential.

Morin's ideas oppose the typecasting of knowledge from disciplines, unless they assume a field of vision that recognizes and conceives the existence of connections and solidarity of their object of study with objects treated by other disciplines, and with the universe of which the object is part (Morin 2000).

4.3 How Does HPM Work?

HPM has attributes that take into account the social, economic and environmental context of farmers, and therefore, unlike IPM, the center of attention is the producers (Fig. 4.1). This means that HPM explicitly recognizes the human being as a component of the system to be studied or managed.

When the objectives for agricultural development are identified, farmers' priorities become essential (Meerman et al. 1996). Therefore, decision-making in HPM is based on axes of sustainable development, the well-being of farmers is sought while respecting nature and reducing negative impacts of pest management towards soci-

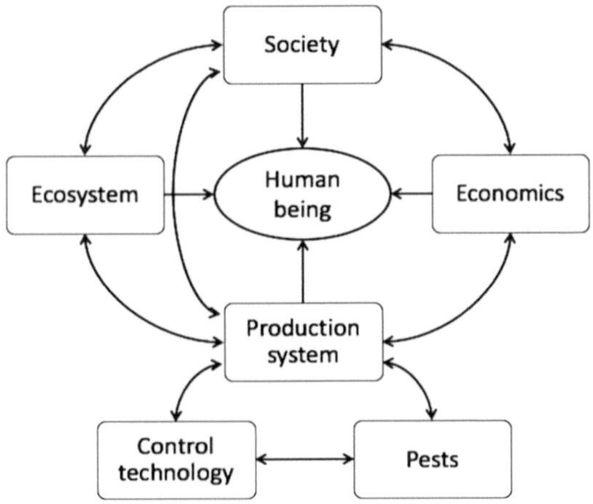

Fig. 4.1 Conceptual diagram showing the components of a Holistic Pest Management program and their relationships. Note that human being – and not pest management – plays the role of articulating axis of the system. (Modified from Barrera 2006)

ety. It is based on the hypothesis that welfare of farmers improves as their agroecosystem is strengthened (or armored) in an ecological, economic and social way.

The premise of HPM starts from the assumption that by overcoming the most significant problems that hinder their daily lives, producers will have the resources and interest to pay attention to pests, enabling them to understand the factors that cause them, and based on that, design healthy production systems. In alternative, when required, they will be capable of developing, acquiring or adopting the technology for pest management that, respecting the environment and taking care of the product quality, achieves competitive levels of profitability.

Thus, the well-being of farmer translates into intellectual, cultural and material resources capable of both improving the resilience of the system and attenuating the frequency and intensity of pest population outbreaks to, finally, decrease or eliminate the need for curative actions.

4.3.1 Concept

As previously stated, HPM is based on two central ideas. First, pest management actions must place the producer at the center of the system; second, pest management should consider not only pests but also other important components of the system in question. HPM is defined as a participatory regional decision-making system in pest management. It is aimed at the welfare of human population. It implements processes and products of low impact and quality for self-consumption and competition in the market. These processes and products are generated from integral production systems. The producer manages these processes as a strategy to focus on the causes of pest outbreaks. Likewise, the producer implements these tactics to minimize the economic, environmental and social costs derived from pest outbreaks and their mismanagement (Table 4.2).

Table 4.2 Fundamentals of Holistic Pest Management (Barrera 2006)

Theme	Definition
Concept	It is a participatory regional system of decision-making in pest management, aimed at the well-being of the human population
Objective	Contribute to regional sustainability
Strategy	Design sustainable agroecosystems
Goal	Manage pests to contribute to the well-being of human population

4.3.2 Objective

HPH aims at contributing to regional sustainability by promoting and strengthening the social, productive and environmental organization of human communities for pest management (Table 4.2).

4.3.3 Strategy

The strategy of HPM is the holistic and participatory design of sustainable agroecosystems through ecological and productive diversification, to prevent and quickly overcome limiting factors such as pest population outbreaks (Table 4.2).

4.3.4 Goal

Pest management must contribute to the well-being of human population through safe and quality processes and products, for self-consumption and positioned in the local, national and international markets (Table 4.2).

4.3.5 Principles

Five principles govern HPM: they are the holistic, participatory, agroecological, safety and fair market principles (Table 4.3).

The holistic principle states that in order to successfully confront pests, it is first necessary to improve the income and welfare of farmers and their families, as well as the capacities that allow them to organize the production and marketing of their products. The participatory principle indicates that the self-management of men and women and the linkage from producer to consumer are necessary to strengthen decision-making and actions in pest management. The agroecological principle proposes are to achieve a more efficient use of internal resources, and a minimum use of external inputs, to favor synergies between processes and components of

Table 4.3 The five principles of Holistic Pest Management (Barrera 2006)

Principle	Definition
Holistic	Improve the income and welfare of farmers and their families, as well as their capacities for the production and marketing of their products
Participatory	Strengthen self-management, decision making and actions in pest management
Agroecological	Promote synergies between processes and components of agroecosystems to prevent pest outbreaks
Innocuousness	Generate safe and quality processes and products
Fair market	Promote fair prices for producers and consumers

agroecosystems that address the causes to prevent pest outbreaks. The innocuousness principle establishes that pest management must generate safe and quality processes and products, both for the ecosystem and for self-consumption and the market. Finally, the equitable market principle indicates that pest management should contribute to favoring remunerative prices for producers and interesting prices for consumers (Table 4.3).

4.4 Pest Control Technology

Figure 4.2 presents the components of pest management and their interactions in the desired context of the holistic approach. It shows that natural control has an important participation in the ecosystem, since through the holistic approach it is intended to conserve and promote the trophic interactions mentioned above, in order to provide the system with mechanisms of "self-protection" or "autonomous control" (Vandermeer et al. 2010).

The health of a crop affected by pests is achieved by maintaining the health of the soil and increasing natural enemies through diversified agroecosystems (Altieri and Nicholls 2003). For this, the agroecosystems must be designed considering four key factors: plant cover (minimum tillage, mulch use, and cover crops), organic matter (manure, compost), nutrient recycling (crop rotation, legume sowing), and pest regulation (introduction and conservation of natural enemies).

Through this strategy, which draws on the concept of ecosystem or environmental services[2] (Costanza et al. 1997), the action of abiotic mortality factors is reinforced by resorting to cultural and soil management practices to conveniently modify the impact of climate on the agroecosystem; it also reinforces the action of biotic mortality factors such as natural enemies (predators, parasitoids and pathogens), through the practice of biological control. This is preferably achieved by conservation, but also by classical biological control since it has shown its benefits

[2]Ecosystem services are the benefits provided to human beings through the transformation of environmental resources or assets (land, water, vegetation and atmosphere) into a flow of goods and services essential for life (clean air, water and food) (Costanza et al. 1997; Department of the Environment, Water, Heritage and the Arts 2009).

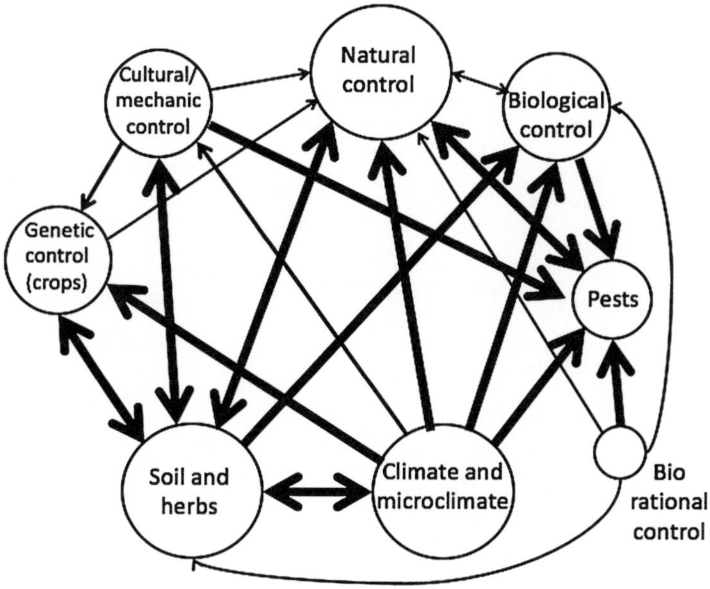

Fig. 4.2 Conceptual diagram of pest control technology under the holistic approach. The size of the circles indicates the relative contribution of a factor or tactic in the system; the thickness of the arrows indicates the influence or impact of a factor or tactic. (Modified from Barrera 2013)

in recovery and ecosystem restoration, especially in the case of invasive plants in wild areas (Headrick and Goeden 2001).

On the contrary, following the agroecological principle of the holistic approach, a strategy based on the use of external inputs is disadvantaged because it promotes the treatment of symptoms. In Fig. 4.2 the biorational control (soft, green or low impact chemicals) is hence represented with a non-absent participation, but very limited.

4.5 Considerations for Developing a Holistic Approach

In addition to understanding the aforementioned concepts, in order to develop pest management in socio-environmental systems under a holistic approach, it is necessary to take into account other elements of the system such as those presented in Fig. 4.3.

From the point of view of pest management with a holistic approach, the elements of Fig. 4.3 can be arranged in actions connected by an ascending spiral, that is, simulating an evolving dynamic system, as presented in Fig. 4.4, what we have called "virtuous circle" of HPM (Barrera 2009). The foundations of each of the actions of this virtuous circle are presented below.

Fig. 4.3 Important elements to develop projects or programs in socio-environmental systems with a holistic approach

4.5.1 The Human Being as the Center of Holistic Thought and Action

The human being is the key articulating element in the socio-environmental systems; therefore, it plays a fundamental role in its dynamics. Thus, under the holistic approach, the human being must be at the center of the socio-environmental system, since the human being can influence the parts of the system to a lesser or greater degree, causing the system to react positively or negatively. Although it is recognized that in nature there are other equally powerful or superior forces, the behavior and attitude of the human being can be modified more easily – comparatively speaking – by means of goodwill and knowledge to reduce negative impacts and promote positive ones (Borrayo 2002).

However, placing the human being at the center of the socio-environmental system entails great responsibility and commitment, since a human being that does not reduce nature only to a further object to meet his needs and utilities is required (Luzzi 2003). Also, to remove the human being from the pragmatic, mechanistic and causal approach that fosters reductionism from the tradition of modern science, an educational action that serves to improve the levels of welfare and social justice is required, with efforts directed to the human being as the subject of his own development (Mata 1994).

Since human actions directly affect socio-environmental systems, and therefore their own development, Borrayo (2002) states that "*the sustainability of all human activity must be a universal human value*". Sustainable Development is based on

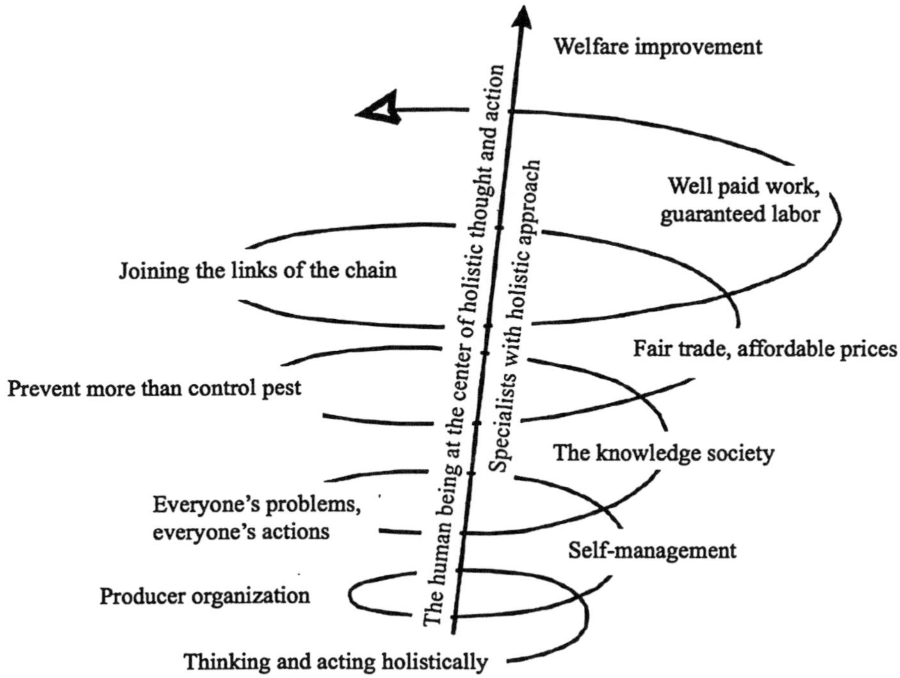

Fig. 4.4 Graphic representation of the HPM virtuous circle. It is an upward spiral that begins by changing the way of thinking and acting (lower part of the figure). By thinking and acting holistically, farmers promote various actions to improve their income and well-being with the support of specialists who facilitate their processes, products and services. This requires promoting organization, participation and self-management in farmers; train them to take advantage of knowledge and technologies to prevent more than control pests; advise farmers in the search for fair price markets and articulate them with the industry to generate added value to their products; all with actions that strengthen attitudes of equity to preserve and improve the workforce. (Modified from Barrera 2009)

this principle, since it is the human being who will choose the development that, meeting the needs of the present, does not compromise the ability for future generations to meet their own needs (WCED 1987).

The poverty of many human communities determines the action and, therefore, the existence of the human being as an individual and as a species. Poor communities or countries contribute significantly to rapid growth of the population and degradation of natural resources, and are a breeding ground conducive to social, economic and environmental instability (Borrayo 2002; FAO 2005). For this reason, poverty has been the object of highest priority attention by all nations of the world (UN 1993). Eliminating poverty – even mitigating it – is a gigantic task, not only because of the immensity of financial resources necessary to ensure food, health and education for the marginalized population, but also for the complexity of the problems it suffers. In this context, any specific action lacking the accompaniment of a holistic approach to achieve a development placing the human being at the center of socio-environmental systems will have minimal chances of success.

4.5.2 First Improve Revenue (Pests Later)

It is practically impossible to effectively manage pests when the poor income of producers allows them only to meet their most basic survival needs. The improvement of income can reverse this situation. For example, Regalado López et al. (1996) found that maize producers in Puebla, Mexico improved their children's food, housing and education when they improved their income by increasing maize production.

Low-income producers do not have access to IPM technology, and when the Government – or non-governmental organizations – provides them training and support for free, in the vast majority of cases the lack of economic resources makes it impossible for them to implement IPM with their own means. Much less, low-income producers will have the resources for the sustainability of IPM in the medium or long term. Therefore, it is not surprising to know that activities of the great mass of projects and programs that support the development of communities culminate when the budget that sustains them is exhausted.

The lack of continuity of pest management projects or programs, beyond the duration of their financing, is largely due to the fact that they do not pursue parallel actions to organize or consolidate the organization of farmers, nor do they seek better markets for their products or to give them added value. Paradoxically, it is very rare that these programs are effectively integrated with those programs or projects that do have those objectives. Invariably, given the dependence of farmers on the State (FAO 2001), they abandon the actions of pest management because they do not have more subsidized resources to give them continuity.

Therefore, as producers' incomes improve, they will have the possibility of allocating certain resources for pest management. Consequently, it is a priority that pest management projects and programs establish the objective of developing the attitudes and skills of producers to improve their income. This is a central objective of the holistic approach (Barrera 2006, 2020).

4.5.3 The Organization of Producers

Producer organization has become a central issue in the discussion of crises in countryside, particularly in developing countries (Fromm and Maccoby 1973; Anonymous 1976; Esteva 1984; Linck 1994; Nelson 1996). And it is not for less, since the disorganization to carry out actions in favor of the collective, and thus individual, well-being, is one of the most important problems of society. Phrases such as "unity is strength" or "together we will win", are almost always demagogic locutions when organizations or groups do not assume the direction of their own development. The lack of power of rural organizations (weak, repressed and/or manipulated organizations), has led non-governmental organizations – mainly – to develop work with groups, with the understanding that the evolution of rural groups will encourage a more complete and equal participation in the process of its development (Muro 1992).

The disorganization of farmers, particularly among the smallest and poorest, makes their production costs more expensive, makes it impossible to sell their products at fair prices and discards them as credit subjects, among other obstacles. Therefore, one of the most important tasks that the holistic approach must have is facilitating legally constituted and functional organizations of producers, and/or consolidating the organizations already established, under the flow of "organization for development" (Anonymous 1976). To achieve this, the development of farmer working groups must be seen as an educational process in itself, aimed at preparing these groups to achieve a more active and more complete inclusion in their development processes (Muro 1992; Mata 1994). This implies, according to Esteva (1984), recognizing, respecting and supporting the organizations that farmers themselves have developed. In this way they move towards higher levels of organization in which a genuine participation in group decisions prevails, allowing them to acquire full control of their future. This will effectively modify their relationship with their political organizations and other social agents and set more ambitious goals. The participation of producers in the transformation of their reality is a key factor in the consolidation of organizations or working groups.

4.5.4 *Self-Management, the Best Inheritance*

It has been recognized that low participation of small-scale coffee producers, and other postponed agricultural groups, has been a limiting factor in achieving sustainable agricultural production and in the proper management of agricultural activities (Galán 1994). For a long time the protectionist development model of Governments in developing countries led to extreme intellectual and financial dependence on farmers (Barajas 2002), although in other cases the dependence was on non-governmental organizations (Pomareda 2001). When the neoliberal development model arrived, and without support and subsidies, the crisis in the countryside broke out loudly due to inability of many farmers to produce profitably and compete successfully in the market (Steffen Riedemann and Echánove Huacuja 2005). According to Linck (1994), when the State withdrew, the "neoliberal shock" accentuated the social differentiation and precariousness of living conditions of the majority of farmers, notably affecting their purchasing power, health and nutritional situation.

The wisest of lessons that these failures have generated has been to recognize that "better than giving fish is teaching to fish". Indeed, developing the capacity of producers to manage their own resources (training, productive projects and markets, among other needs), is the most important mechanism of the holistic approach. Self-management, that is the activity that allows rural and city workers, through their organizations, to (i) take into their hands (and without intermediaries or agents external to the organization), all the problems that arise in their respective groupings and (ii) manage, by themselves, the tasks and actions necessary for the benefit of the organization, has been indicated, along with the participatory methodology, as the only real way for farmers to improve their living conditions and seek own development alternatives (Mata 1994).

The type of model generating the appropriate pest management technology influences the farmers' self-management capabilities. Nelson (1996) found that a model of technology generation guided by farmers obtained a greater number of self-management process levels (6 out of 7 levels) than a model guided by scientists (2 out of 7).

4.5.5 Participation: Problems of All, Actions by All

The first objective of a comprehensive study is to obtain a diagnosis of how the system works (García 1994). In the context of a holistic approach, it is a priority to make the diagnosis of the reality with the participation of the society, to detect the priority problems that afflict it (Barrera 2006, 2020). Participatory diagnosis will not only generate the baseline on which future work will rely, but will also provide the first opportunity to launch the holistic approach by immersing in the complexity of the system's problem and finding the interactions of its elements.

In addition, the identification of the system problems by the affected society itself will guarantee the application of available resources to its most felt problems. In this way, any achievement in solving the problems will have the greatest impact on the system.

Planning the actions to solve the detected problems, like the diagnostic phase, must be a participatory task. The farmers define what to do, how, where, when and by whom. They are the ones who perceive the problems and those who will benefit from the actions they define. Since the 1970s it is clear that "*agricultural and support policies cannot and should not be designed from above*", (Warman 1979), while Altieri and Masera (1993) indicated that conventional strategies "from above" have been unable to promote sustainable development in Latin America.

Therefore, participation is not only a means of improving production, it is rather the objective to be achieved (Matteson 2000). Thus, the analysis of the problem, the definition of the activities and the monitoring processes must be participatory tasks. The holistic approach is greatly enriched by the participation of the affected society.

4.5.6 Knowledge Society

It is said that a "knowledge society" is one that has the capacity to generate, appropriate and use knowledge in order to meet the needs of its development. Today, more than ever, it is accepted that scientific knowledge and technological innovations contribute significantly to the development of society (Parra 2004). The knowledge generated through research is transformed into information and this into wealth. Technological innovations make the difference between producing more and with higher quality at lower costs, essential requirements to compete successfully in the market. However, to facilitate the adoption of technologies by farmers it is important that they participate from the outset in the research processes through participatory and integrative mechanisms (Pareja 1992).

It should be stressed, however, that the knowledge society to which we refer from the perspective of HPM is that whose members can appropriate and use the available knowledge, that is scientific, technological as well as traditional. They can generate knowledge to better understand their problems, and from it propose solutions and take actions to solve them (Olivé 2009).

Farmers can generate their own knowledge and technologies or acquire them through training and technical advice (e.g., van der Berg et al. 1997; Chikoye et al. 1999; Sibanda et al. 2000; Joshi et al. 2000; Phiri et al. 2001; Songa et al. 2002; Segura et al. 2004). Under the holistic approach, it is a priority to encourage farmers to be trained and participate in continuous training processes. Likewise, local capacities must be developed through the formation of technical cadres and community promoters. In this sense, the Farmer Field Schools (FFS) are a good example to develop local training and advisory capabilities under the principle of "first farmer" and based on learning with practice and participatory research (Braun et al. 2000; Matteson 2000; Jarquín 2003; Jarquín et al. 2018).

4.5.7 Agricultural Production and Conservation of Natural Resources

With respect to pesticide contamination, there is a growing concern and resistance of consumers for the excessive use of these products in food production (Wearing 1988; Hollingsworth and Coli 2004). Therefore, producing healthy and sufficient food for self-consumption and the market, through agroecological actions, is vital under the holistic approach in order to conserve natural resources and strengthen natural processes or cycles (Barrera 2006, 2020). Organic fertilizers to improve soil fertility, terraces or live barriers to prevent erosion, diversification of productive systems to promote biodiversity, improve soils, retain water and promote natural pest control, as well as the use of biological control agents, natural insecticides, traps with attractants, resistant crop varieties, etc., are some of the methods and tools available for agricultural production and conservation of natural resources.

Reducing external inputs, recycling crop residues, avoiding pollution and making use of alternative energy sources, all objects of agroecology (Altieri 1987), are components of the holistic approach, where a harmonious interaction among society, nature and economy is encouraged.

These actions must be accompanied, in the agroecological management of production systems, by projects for payment for environmental services (Costanza et al. 1997; Anta 2006; Gómez-Baggethun et al. 2010) and ecotourism (Carballo 1998; Scheyvens 1999; Weaver 2001), among others, related to obtain economic resources through the conservation of natural resources in rural communities. The holistic approach thus develops awareness of environmental conservation, as a generator of wealth and well-being.

4.5.8 *Preventing Is Better Than Curing*

Among farmers, the use of agrochemicals to solve various crop problems such as low soil fertility or pest infestation is very entrenched. Without the "liquids" (agrochemicals or other inputs), they feel unprotected or inert, even when their application is not justified or solved the root problems.

Under the holistic approach, the attention of many of these problems requires substantive or "ecological engineering" actions (Gurr et al. 2004), that is propitiating the conditions to treat their causes, not their symptoms (Dreves 1996). Thus, it is privileged to conserve and improve soil and biodiversity of the agroecosystem in order to obtain more tolerant crops with a greater capacity for stress recovery, but also to create environments that favor natural cycles (water, carbon, nitrogen) and natural pest control (Altieri 1994; Altieri and Nicholls 2003; Vandermeer et al. 2010). The holistic approach is intended to ecologically "shield" the agroecosystem against abiotic and biotic adversities, which leads to follow a strategy of pest prevention rather than control.

4.5.9 *Fair Trade, Affordable Prices*

In the conventional market the success of a farmer is measured by his net profit. If the producer spends more than he finally obtains from the sale of his crop, then he loses and fails, while if the profit far exceeds the costs of production, the business thrives and success – in terms of satisfaction of needs, savings and investment – will be assured. Unfortunately, low prices often affect producers' incomes, especially if their product is not sold in special markets – e.g., organic products (Willer and Yussefi 2005) – nor does it have an added value.

Under a holistic approach, it is intended to promote or access markets whose prices reward the efforts of farmers and recognize the environmental and social services they provide to local and global society. Under this approach it is also recognized that prices must be affordable to consumers. Farmers have the right to fair payment for their products, as well as the society has right at affordable prices for the goods it requires (i.e. food, health, clothing or housing) (Vanderhoff 2005).

Farmers, society and government must deploy every effort to develop alternative markets under the holistic approach. At the local level, organic small markets (Bustamante-Lara and Schwentesius-Rindermann 2018) with participatory certification (Astorga 2003) have great potential, as they offer a space for the flourishing of a direct relationship (without intermediaries) between producers and consumers.

4.5.10 Joining the Links in the Chain

The disarticulation of the industrial production process is one of the most common problems in the agricultural sector and an opportunity missed by many farmers (FAO 2014). Since transforming raw materials into processed goods can significantly improve farmers' incomes (Sebatta et al. 2015; Gashaw et al. 2018), the industrialization of products should be a holistic approach priority to give them added value.

Entering into the industrialization of their products means for farmers a great effort that only consolidated organizations can assume (Barrera et al. 2004). Therefore, under a holistic approach, emphasis should be placed on improving the managerial, economic and technical capacities of consolidating organizations, especially developing an entrepreneurial attitude among farmers.

4.5.11 Well-Paid Work, Insured Labor

If farmers deserve fair prices for their products, also farm workers or laborers are entitled to a fair remuneration for their services. Farmworkers or day laborers that often are members of whole families, are possibly among the poorest wage earners in the economically active labor sector (Krumbiegel et al. 2018; Moyce and Schenker 2018). However, they are rarely treated well (food, housing, education and medical insurance). The socio-environmental system will never achieve sustainability if there is such social inequity.

It is not surprising, then, that labor is scarce and that its cost is constantly increasing. The labor market suffers very strong imbalances when product prices are low, among other causes (Martin 2016). It is hence very important under the holistic approach to develop or grant access to markets that pay fair prices, but also to guarantee decent treatment of agricultural workers. This concern of "going beyond what is produced to deal with how it is produced", had its peak in the 1990s under the concept of "Corporate or Corporate Social Responsibility", through which many companies adopted codes of conduct on social and environmental aspects (Jenkins 2004).

4.5.12 An Example

The points developed in the previous sections have been organized in Fig. 4.5 to give an example that summarizes objectives, strategies and components of an HPM program. The objectives pursued in this example are to improve income, cultivate while conserving natural resources and distribute benefits equally.

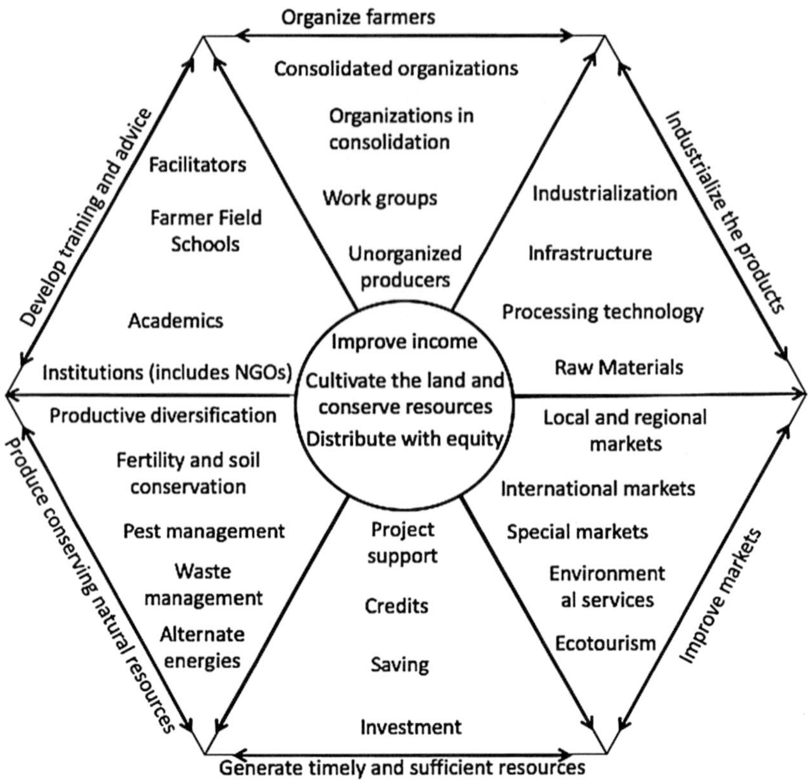

Fig. 4.5 Objectives (center) strategies (sides of the hexagon) and main components (inside the triangles) of a HPM program. It considers carrying out actions on different elements of the system to facilitate the management of pests. (Modified from Barrera 2009)

To meet these objectives, the following six strategies are proposed: organize the producers; industrialize the products; improve markets; generate timely and sufficient resources; produce agricultural products and conserving natural resources; and develop training and local technical advice.

4.6 The Risk as a Cornerstone of the HPM

Two interconnected concepts emerge when operating a system through HPM (Barrera et al. 2007):

(i) Reduce risk and vulnerability, and
(ii) Increase response capacity and resilience.

In HPM, the "risk" is understood as the possibility that farmers may suffer damage due to a pest, while "vulnerability" includes the characteristics or proprieties of farmers and farms, which determine their degree of exposure to a pest. For its part,

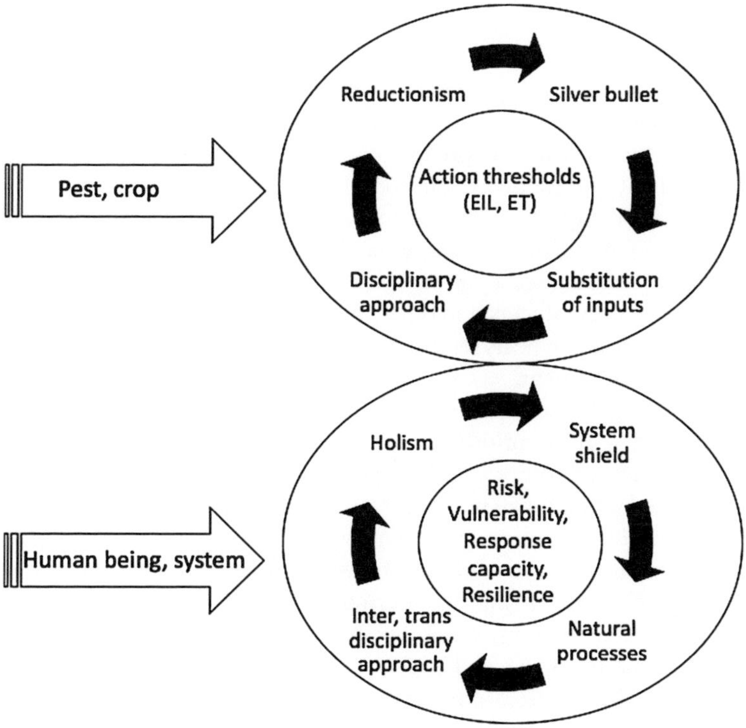

Fig. 4.6 Contrast in decision making between IPM and HPM. In the case of IPM (upper circle), decision making is based on action thresholds: Economic Injury Level (EIL) and Economic Threshold (ET). In HPM, decisions are oriented based on risk, vulnerability and system resilience. Contrary to action thresholds that governs pest management under the disciplinary and reductionist approach of IPM, risk, vulnerability and resilience govern pest management under the transdisciplinary and holistic approach of HPM. (Modified from Barrera et al. 2007)

"resilience" is the ability of households, communities and nations to absorb and recover from impacts, while positively adapting their structures and means to live under stress, or changes and uncertainties, in long-term periods (Mitchel 2013). These concepts are related to response and recovery capacity of the system, which are the attributes and mechanisms of producers, organizations, institutions and society to reduce risks of pests and to survive, resist and recover from the damage caused by them. By reducing risk and vulnerability and increasing responsiveness and resilience, it is intended to "shield" the system ecologically, economically and socially.

These concepts are the cornerstone of HPM, because not only is it important to reduce pest incidence but also to reduce vulnerability, improving the response and resilience capabilities of the system under management.

Contrary to Economic Injury Level that is the result of a disciplinary approach of IPM, the duet formed by risk and vulnerability on the one hand, and response capacity and resilience on the other, gives rise to an approach inter- and transdisciplinary that is only achieved under HPM (Fig. 4.6).

4.7 Conclusion

HPM is presented as the new paradigm for protection of plants, animals and human beings. This paradigm is nourished by the convergence of three schools of thought: holistic management (system approach), agroecology (agricultural production and conservation of natural resources) and complex thinking (emphasis on interactions). Based on the well-being and empowerment of the human being, HPM promotes the construction of a participatory and regional decision-making scaffolding. It seeks at minimizing problems caused by pests by reducing vulnerability and increasing the response capacity of producers and other links in the productive agricultural chain. HPM reduces the risk that pests represent and increases the resilience of the systems under study or management.

Before using curative methods for pest management – such as chemical or biological pesticides – HPM prioritizes preventive methods. To achieve this, HPM relies on the design of comprehensive agroecological production systems. In this way, it is intended to minimize pest outbreaks by promoting natural pest control and using curative methods as a last option. On the other hand, systems managed through HPM must guarantee food sovereignty and well-paid work for agricultural workers.

All this with the support and monitoring of holistic teams (producers-trainers-academics) with the objective of producing safe and value-added processes and products, both for the local and global market, whose dividends contribute to improving the quality of life of the farmers' families.

References

Altieri, M. A. (1987). *Agroecology: The scientific basis of alternative agriculture*. Boulder: Westview Press.

Altieri, M. A. (1994). *Biodiversity and pest management in agroecosystems*. New York: Hayworth Press.

Altieri, M. A., & Masera, O. (1993). Sustainable rural development in Latin America: Building from the bottom-up. *Ecological Economics, 7*(2), 93–121.

Altieri, M. A., & Nicholls, C. I. (2003). Ecologically based pest management: A key pathway to achieving agroecosystem health. In D. J. Rappoport, W. L. Lasley, D. E. Rolston, et al. (Eds.), *Managing for healthy ecosystems* (pp. 999–1010). Boca Raton: Lewis Publishers.

Anonymous. (1976). *Los campesinos ¿para que organizarlos?* (1st ed.). Mexico: Centro de Ecodesarrollo. Consejo Nacional de Ciencia y Tecnología, 212 pp.

Anta, S. (2006). El café de sombra: Un ejemplo de pago de servicios ambientales para proteger la biodiversidad. *Gaceta Ecológica, 80*, 19–31.

Astorga, A. (2003). Curso-Taller Gestión de mercados locales en América Latina. 27 de agosto al 1 de septiembre, 2003. Corporación Educativa para el Desarrollo. Costa Rica.

Avelino, J., Cristancho, M., Georgiou, S., et al. (2015). The coffee rust crises in Colombia and Central America (2008–2013): Impacts, plausible causes and proposed solutions. *Food Security, 7*, 303–321.

Barajas, G. (2002). Las políticas de atención a la pobreza en México, 1970–2001: De populistas a neoliberales. *Revista Venezolana de Gerencia, 7*, 553–578.

Barrera, J. F. (2006). Manejo holístico de plagas: Hacia un nuevo paradigma de la protección fitosanitaria. In J. Pohlan, L. Soto, & J. Barrera (Eds.), *El cafetal del futuro: Realidades y Visiones* (pp. 63–82). Aachen: Shaker Verlag.

Barrera, J. F. (2009). La necesidad del enfoque holístico en el manejo de plagas. In J. C. Arrivillaga, M. El Souki, & B. Herrera (Eds.), *Enfoques y temáticas en entomología* (XXI Congreso Venezolano de Entomología) (pp. 1–28). Caracas: Sociedad Venezolana de Entomología.

Barrera, J. F. (2013). El control biológico bajo el enfoque del manejo holístico de plagas. In M. C. del Rincón Castro & V. Ríos Olivera (Eds.), *Memorias del XXIV Curso Nacional de Control Biológico* (pp. 182–193). Sociedad Mexicana de Control Biológico y Centro Interdisciplinario de Investigación para el Desarrollo Integral Regional (CIIDIR), Unidad Oaxaca del Instituto Politécnico Nacional. Oaxaca de Juárez, Oaxaca, Mexico, 4–6 November, 2013.

Barrera, J. F. (2020). Holistic Pest Management. In D. Perez-Staples, F. Diaz-Fleischer, P. Montoya, & M. T. Vera (Eds.), *Area-wide management of fruit fly pests* (pp. 275–293). Boca Raton: CRC Press, Taylor & Francis Group.

Barrera, J. F., Parra Vázquez, M., Herrera Hernández, O. B., et al. (2004). *Plan Estatal de Manejo Agroecológico del Café en Chiapas: Guía hacia una cafeticultura sustentable*. Tapachula: Comisión para el Desarrollo y Fomento del Café de Chiapas y El Colegio de la Frontera Sur, 164 pp.

Barrera, J. F., Herrera, J., & Gómez, J. (2007). Riesgo-vulnerabilidad hacia la broca del café bajo un enfoque de manejo holístico. In J. F. Barrera, A. García, V. Domínguez, & C. Luna (Eds.), *La Broca del Café en América Tropical: Hallazgos y Enfoques* (pp. 131–141). Mexico: Sociedad Mexicana de Entomología y El Colegio de la Frontera Sur.

Barrera, J. F., Avelino, J., Huerta, G., et al. (2013). La roya del café: Crónica de una devastación anunciada. *Ecofronteras, 49*, 22–25.

Borrayo, R. (2002). *Sustentabilidad y desarrollo económico*. México: McGraw-Hill, 124 pp.

Braun, A. R., Thiele, G., & Fernandez, M. (2000). *Farmer Field Schools and Local Agricultural Research Committees: Complementary platforms for integrated decision-making in sustainable agriculture*. London: Agricultural Research and Extension Network, ODI.. AgREN Network Paper No. 105.

Bruil, J., Anderson, C., Bernhart, A., et al. (2019). *Strengthening FAO's commitment to agroecology*. Coventry: Centre for Agroecology, Water and Resilience. Coventry University.

Bustamante-Lara, T. I., & Schwentesius-Rindermann, R. (2018). Profile and situation of producers who form the street and organic markets in Mexico. *Agricultura, Sociedad y Desarrollo, 15*, 507–530.

Carballo, A. (1998). Concepción y perspectivas del ecoturismo en México. *Planeta Mexico Bulletin*. http://www.planeta.com/planeta/98/0898ecotur.html. Accessed 16 Sept 2019.

Chikoye, D., Ekeleme, F., & Ambe, J. T. (1999). Survey of distribution and farmers' perceptions of speargrass [*Imperata cylindrica* (L.) Raeuschel] in cassava-based systems in West Africa. *The International Journal of Pest Management, 45*(4), 305–311.

Costanza, R., d'Arge, R., de Groot, R., et al. (1997). The values of the world's ecosystem services and natural capital. *Nature, 387*, 253–260.

Department of the Environment, Water, Heritage and the Arts. (2009). *Ecosystem services: Key concepts and applications* (Occasional Paper No 1). Canberra: Department of the Environment, Water, Heritage and the Arts.

Dreves, A. J. (1996). Village-level Integrated Pest Management in developing countries. *Journal of Agricultural Entomology, 13*(3), 195–211.

Ehler, L. E., & Bottrell, D. G. (2000). *The illusion of Integrated Pest Management*. Issues Sci Technol. http://www.issues.org/16.3/index.html. Accessed 28 June 2019.

Esteva, G. (1984). *La batalla en el México rural* (4ª ed., p. 243). Mexico: Siglo XXI Editores.

FAO (Organización de las Naciones Unidas para la Agricultura y la Alimentación). (2001). *El estado mundial de la agricultura y la alimentación 2001*. Roma, 295 pp.

FAO (Organización de las Naciones Unidas para la Agricultura y la Alimentación). (2005). *El estado de la inseguridad alimentaria en el mundo 2005. La erradicación del hambre en el mundo: Clave para la consecución de los objetivos de desarrollo del Milenio*. Roma, 36 pp.
FAO (Organización de las Naciones Unidas para la Agricultura y la Alimentación). (2014). *Developing sustainable food value chains – Guiding principles*. Rome.
Fromm, M., & Maccoby, M. (1973). *Socio-psicoanálisis del campesino mexicano: Estudio de la economía y la psicología de una comunidad rural* (p. 395). Mexico: Fondo de Cultura Económica.
Galán, B. B. (1994). *Participación campesina para una agricultura sostenible en países de América Latina. Participación Popular 7*. Roma: Organización de las Naciones Unidas para la Agricultura y la Alimentación.
García, R. (1994). Interdisciplinariedad y sistemas complejos. In E. Leff, R. García, P. Gutman, V. M. Toledo, H. M. C. Vessuri, R. Fernández, & R. Brañes (Eds.), *Ciencias Sociales y Formación Ambiental* (pp. 85–124). Mexico: Gedisa Editorial.
Gashaw, B. A., Habteyesus, D. G., & Nedjo, Z. S. (2018). Determinants of coffee value addition by smallholder farmers in Jimma Zone, Ethiopia. *The International Journal of Business Management and Technology, 2*(4), 112–123.
Gliessman, S. R. (2002). *Agroecología: Procesos ecológicos en agricultura sostenible*. Turrialba: LITOCAT.
Gómez-Baggethun, E., de Groot, R., Lomas, P. L., et al. (2010). The history of ecosystem services in economic theory and practice: From early notions to markets and payment schemes. *Ecological Economics, 69*, 1209–1218.
Gurr, G. M., Wratten, S. D., & Altieri, M. A. (2004). *Ecological engineering for pest management. Advances in habitat manipulation for arthropods*. New York: Comstock Publishing Associates, 232 p.
Headrick, D. H., & Goeden, R. D. (2001). Biological control as a tool for ecosystem management. *Biological Control, 21*, 249–257.
Hollingsworth, C. S., & Coli, W. M. (2004). Consumer response to IPM: Potential and challenges. In O. Koul, G. S. Dhaliwal, & G. W. Cuperus (Eds.), *Integrated Pest Management: Potential, constraints and challenges* (pp. 255–264). Wallingford: CAB International.
Jarquín, R. (2003). Las ECEAs: Base para la implementación de proyectos de desarrollo autogestionarios en zonas cafetaleras. *LEISA Revista de Agroecología, 19*, 33–36.
Jarquín, R., Ramírez Tobías, H. M., Aguilar Benítez, G., et al. (2018). *Las Escuelas de Campo y Experimentación para Agricultores (ECEA's): Guía metodológica y logros alcanzados en Chiapas y San Luis Potosí*. Mexico: Universidad Autónoma de San Luis Potosí.
Jarquín-Gálvez, R., Jiménez Sánchez, L., Guharay, F., et al. (2005). Aportes de productores y científicos al entendimiento de la agregación de *Hypothenemus hampei* en Chiapas, México. *Manejo Integrado de Plagas y Agroecología, 75*, 77–82.
Jarquín-Gálvez, R., Barrera, J. F., García-Barrios, L., et al. (2006). Manejo Integrado de *Hypothenemus hampei* bajo dos modelos de capacitación y difusión en México. *Manejo Integrado de Plagas y Agroecología, 78*, 36–45.
Jenkins, R. (2004). Códigos de conducta empresariales: Autorregulación en una economía global. *Comercio Exterior, 54*, 764–778.
Jiménez, L., Barrera, J. F., Rojas, M., et al. (2000). Tácticas y estrategias usadas por ejidatarios para combatir a la "Broca del café" *Hypothenemus hampei* (Coleoptera: Scolytidae), en el municipio de Tapachula, Chiapas. In *XXIII Congreso Nacional de Control Biológico* (pp. 208–210). Guanajuato: Guanajuato.
Joshi, R. C., Matchoc, O. R. O., Bahatan, R. G., et al. (2000). Farmers' knowledge, attitudes and practices of rice crop and pest management at Ifugao Rice Terraces, Philippines. *The International Journal of Pest Management, 46*(1), 43–48.
Krumbiegel, K., Maertens, M., & Wollni, M. (2018). The role of fairtrade certification for wages and job satisfaction of plantation workers. *World Development, 102*, 195–212.

Linck, T. (1994). Síntesis y comentarios. In T. Linck (Ed.), *Agriculturas y campesinados de América Latina. Mutaciones y recomposiciones. Institut Français de Recherche Scientifique pour le Développement en Coopération* (pp. 31–34). Mexico: Fondo de Cultura Económica.
Luzzi, D. (2003). La "ambientalización" de la educación formal. Un diálogo abierto en la complejidad del campo educativo. In E. Leff (Coord), *La complejidad ambiental* (2nd ed., pp. 158–192). México: Siglo XXI Editores, SA de CV.
Martin, P. L. (2016). *Migrant workers in commercial agriculture*. Geneva: International Labour Organization.
Mata, B. (1994). *Un modelo participativo y autogestivo de educación campesina* (2da ed., p. 86). México: Departamento de Sociología Rural, Universidad Autónoma Chapingo.
Matteson, P. C. (2000). Insect pest management in tropical Asian irrigated rice. *Annual Review of Entomology, 45*, 549–574.
Meerman, F., Van De Ven, G. W. J., Van Keulen, H., et al. (1996). Integrated crop management: An approach to sustainable agricultural development. *The International Journal of Pest Management, 42*(1), 13–24.
Mitchell, A. (2013). *Risk and resilience: From good idea to good practice: A scoping study for the Experts Group on Risk and Resilience*. Paris: Organisation for Economic Cooperation and Development (OECD), Development Co-operation Working Paper
Morin, E. (2000). *La mente bien ordenada: Pensar la reforma, reformar el pensamiento* (p. 185). Barcelona: Ed. Siex Barral SA.
Morin, E. (2006). *El Método* (7a ed.). Madrid: Ediciones Cátedra, 448 pp.
Morse, S. (2009). IPM: Ideals and realities in developing countries. In E. B. Radcliffe, W. D. Hutchinson, & R. E. Cancelado (Eds.), *Integrated Pest Management: Concepts, tactics, strategies and case studies* (pp. 458–470). Cambridge: Cambridge University Press.
Morse, S., & Buhler, W. (1997). *Integrated pest management: Ideals and realities in developing countries*. London: Lynne Rienner Publishers.
Moyce, S. C., & Schenker, M. (2018). Migrant workers and their occupational health and safety. *Annual Review of Public Health, 39*, 351–365.
Muro, P. (1992). *Problemas del campesinado y desarrollo rural alternativo. Ensayos 1981–1987.* Mexico: Universidad Autónoma Chapingo, 186 pp.
Nelson, K. C. (1996). Estudio comparativo de la generación de tecnología MIP en el cultivo de tomate, Nicaragua. *Manejo Integrado de Plagas, 41*, 16–28.
Olivé, L. (2009). Por una auténtica interculturalidad basada en el reconocimiento de la pluralidad epistemológica. In E. Sader (Ed.), *Pluralismo epistemológico. Consejo Latinoamericano de Ciencias Sociales (CLACSO)* (pp. 19–30). La Paz: Muela del Diablo Editores.
Pareja, M. R. (1992). El Manejo Integrado de Plagas: Componente esencial de los sistemas agrícolas sostenibles. *Manejo Integrado de Plagas, 24–25*, 44–50.
Parra, M. R. (2004). Vinculación: Gestión de la innovación basada en el conocimiento. *Ecofronteras, 23*, 24–28.
Parsa, S., Morse, S., Bonifacio, A., et al. (2014). Obstacles to integrated pest management adoption in developing countries. *Proceedings of the National Academy of Sciences of the United States of America, 111*(10), 3889–3894.
Phiri, N. A., Hillocks, R. J., & Jeffries, P. (2001). Incidence and severity of coffee diseases in smallholder plantations in northern Malawi. *Crop Protection, 20*, 325–332.
Pomareda, C. (2001). Los pequeños productores y su participación en las agroexportaciones en Centroamérica. In *El sector agroalimentario: Integración regional y vinculaciones internacionales para su desarrollo.* San Isidro de Coronado: Taller Regional de la UNCTAD-IICA-CORECA-CAC.
Regalado López, J., Jiménez Sánchez, L., Laird, R. J., et al. (1996). Factores asociados a la utilización de la tecnología de alta productividad entre productores de maíz de temporal. *Agrociencia, 30*(1), 139–145.
Savory, A., & Butterfield, J. (1999). *Holistic management, a new framework for decision making* (2nd ed.). Washington, DC: Island Press, 616 pp.

Savory Global. (2019). *A thriving community working to advance holistic management*. https://www.savory.global/our-mission/. Accessed 9 Sept 2019.

Scheyvens, R. (1999). Ecotourism and the empowerment of local communities. *Tourism Management, 20*, 245–249.

Sebatta, C., Mugisha, J., Katungi, E., et al. (2015). Adding value at the farm: The case of smallholder potato farmers in the Highlands of Uganda. *Asian Journal of Agricultural Extension, Economics & Sociology, 4*(3), 210–223.

Segura, H. R., Barrera, J. F., Nazar, A., et al. (2004). Farmers' perceptions, knowledge and management of the coffee pests and diseases and their natural enemies in Chiapas, Mexico. *Journal of Economic Entomology, 97*, 1491–1499.

Sibanda, T., Dobson, H. M., Cooper, J. F., et al. (2000). Pest management challenges for smallholder vegetable farmers in Zimbabwe. *Crop Protection, 19*, 807–815.

Songa, J. M., Overholt, W. A., Mueke, J. M., et al. (2002). Farmers' perceptions of aspects of maize production systems and pests in semi-arid eastern Kenya: Factors influencing occurrence and control of stemborers. *The International Journal of Pest Management, 48*(1), 1–11.

Steffen Riedemann, C., & Echánove Huacuja, F. E. (2005). El modelo neoliberal y el difícil proceso organizativo que viven los ejidatarios mexicanos productores de granos. *Polis: Investigación y Análisis Sociopolítico y Psicosocial, 1*, 211–233.

UN (United Nations). (1993). *Agenda 21: Earth Summit – The United Nations Programme of Action from Rio*, 294 pp.

van der Berg, H., Ankasah, D., Muhammad, A., et al. (1997). Evaluating the role of predation in population fluctuations of the soybean aphid *Aphis glycines* in farmers' fields in Indonesia. *Journal of Applied Ecology, 34*, 971–984.

Vanderhoff, F. (2005). *Excluidos hoy, protagonistas mañana*. Mexico: Impretei SA de CV, 166 p.

Vandermeer, J., Perfecto, I., & Philpott, S. M. (2010). Ecological complexity and pest control in organic coffee production: Uncovering an autonomous ecosystem service. *Bioscience, 60*, 527–537.

Warman, A. (1979). Andamos arando. El problema agrario campesino (1978). *Nexos, 13*, 11–16.

WCED (World Commission on Environment and Development). (1987). Our common future: Report of The World Commission on Environment and Development. General Assembly, 4 August 1987, United Nations, 374 pp.

Wearing, C. H. (1988). Evaluating the IPM implementation process. *Annual Review of Entomology, 33*, 17–38.

Weaver, D. B. (Ed.). (2001). *The encyclopedia of ecotourism*. Wallingford: CAB International.

Willer, H., & Yussefi, M. (Eds.). (2005). *The world of organic agriculture. Statistics and emerging trends 2005* (7th ed.). Bonn: International Federation of Organic Agriculture Movements, 189 pp.

Chapter 5
Methods for Analysis of Holistic Pest Management

... The important thing is not the methods, but the use of certain tools in a given research framework. That is, ask the right questions and look for methods that are appropriate for them and for the context.
Ian Scoones, Sustainable livelihoods and rural development (Translated from: Scoones I (2017) Medios de vida sostenibles y desarrollo rural. Icaria Editorial, S.A. Barcelona, p. 147).

Abstract In this chapter some tools or methods to generate information on the organization, structure and operation of socio-environmental systems that require the implementation of Holistic Pest Management (HPM) are presented. They are discussed in an illustrative, non-prescriptive way. The chapter presents a combination of methods that proceed from the social sciences such as the Sustainable rural livelihoods and the structural analysis, with others coming from the natural sciences such as the Rapid Agroecological Sampling. The description of these methods is organized into the following topics: (i) Integration and training of the work team, (ii) Identification of limits and interactions between system components, (iii) Preparation of the work plan and, (iv) Development of decision-making and monitoring systems. To develop and execute the work plan, the importance of an interdisciplinary team with producers, facilitators and academics participation is emphasized. Likewise, it is advised that this working group has to be trained following the methodology of Farmer Field Schools and Agricultural Experimentation. The methodology of the Holistic Risk Index (HRI), considered the cornerstone of the HPM, is described. Since the HPM ultimate goal is to increase the resilience of the socio-environmental system, a conceptual approach is presented that relates risk (HRI) to an approach to measure resilience.

Keywords Sustainable rural livelihoods · Rapid agroecological sampling · Farmer field schools · Holistic risk index · Resilience

J. F. Barrera, *Beyond IPM: Introduction to the Theory of Holistic Pest Management*, Sustainability in Plant and Crop Protection 15,
https://doi.org/10.1007/978-3-030-43370-3_5

5.1 Introduction

Unlike the Integrated Pest Management (IPM) that focuses on assessing the effects of pests on crops and, from there, making management decisions, the Holistic Pest Management (HPM) requires – in addition to considering the effect of pests – detailed information on the organization, structure and operation of the socio-environmental system to study or manage.

HPM implementation requires a conceptual and methodological framework that contains, as suggested by Montes and Leff (1986), "*the different elements, variables and processes that affect a specific problem field, that evaluates the specific weight that they adopt within a complex system of interrelationships, and that guides the actions leading to their resolution*". In Chap. 4 the conceptual framework of the HPM was approached, now, in Chap. 5 we will focus on the methodological framework.

The first step in analyzing a socio-environmental system is to identify its components and limits (Hart 1985; García 1994). The information generated by this analysis allows the understanding of the system structure and function, and from there, implementing the HPM to reduce the risk towards pests and to increase the system resilience. Several methods are followed in order to implement an HPM program, such as those shown in Table 5.1.

Although the means (methods) may vary depending on circumstances, the purpose (goal) must be maintained and guided by the following steps:

(i) Integration and training of the work team
(ii) Identification of limits and interactions among system components
(iii) Preparation of the work plan
(iv) Development of decision-making and monitoring systems
(v) Implementation of the work plan

Table 5.1 Summary of methodologies for implementing the Holistic Pest Management

Concept	Objective	Method
Work team	Integrate and train the interdisciplinary working group	Farmer field schools and agricultural experimentation
Diagnosis	Identify the limits and main components or elements of the system	Brainstorming Prioritization of problems
	Determine the interactions among system components	Structural analysis
Planning	Define objectives, strategies, actions and goals of the work plan	Planning techniques
Decisions-making	Define the decision-making and monitoring systems	Rapid Agroecological sampling holistic risk index
Implementation	Implement the work plan	Agroecological management techniques

Based on the steps mentioned above and with exception of step v) (Implementation of the work plan) that will be developed in Chap. 6, this chapter describes the main analysis methods used in HPM to make pest management decisions in socio-environmental systems.

5.2 Integration and Training of the Work Team

5.2.1 *The Work Team*

Integrating a multidisciplinary working group is very important to carry out the HPM program. As shown in Fig. 5.1, this group will identify the problem of the socio-environmental system through a diagnosis; plan and execute system management actions; and define the procedure for decision-making. Purpose is orienting the system in question towards the objectives and goals defined by the group (Barrera 2006, 2020; Barrera et al. 2007, 2018).

The composition of the group members is also important; preferably, the team should be made up of the following three subgroups: producers, academics and at least one facilitator or extension agent.

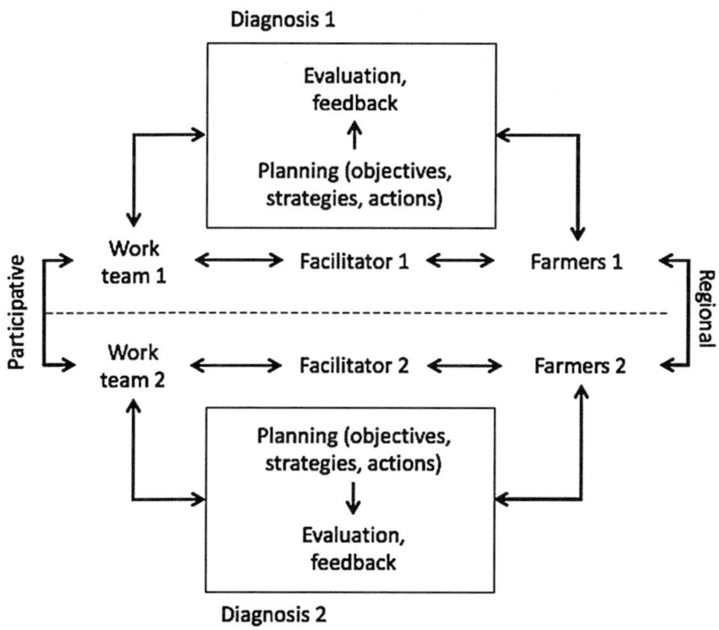

Fig. 5.1 Diagram showing the actions and interaction between the work teams that carry out participatory diagnoses and planning, as well as the regional evaluation of the HPM program of a socio-environmental system. (Modified from Barrera et al. 2007)

5.2.1.1 Producers

Producers are the owners of the productive systems and the main stakeholders interested in improving income and solving pest problems. It is suggested that this subgroup be made up of leading producers, community promoters or agricultural technicians representing different communities or producer organizations; these people will receive the training and, subsequently, they will have the responsibility of training producers in their communities. Ideally, each subgroup of producers should be made up of a maximum of 30 people to facilitate their operation.

5.2.1.2 Academics

This subgroup is made up of staff from academic institutions, preferably specialists in different fields of natural, social and exact sciences. According to Reichlderfer (1980), good communication between specialists is essential for the success of research on pest management and other agricultural issues. The value of specialist participation from different disciplines is that they provide the overall vision for crop management and enrich the understanding of phytosanitary problems (Hilje and Ramírez 1992).

Academics have the responsibility of providing objective information about the various phenomena that occur in socio-environmental systems that help to understand their organization, structure and operation. They must also propose solutions to problems that are practical and conveniently adapted to the producers circumstances. An academic with holistic approach is a professional with broad and deep skills in a discipline, who is able to understand the importance of other specialists' disciplines and the need to interact with other team members to solve complex problems.

5.2.1.3 Facilitators

The facilitators are in charge of two functions: to serve as a liaison between producers and academics; and to follow up so that the HPM program is carried out in a timely manner. Among the capabilities that these people should have there are those of "translating" the scientific-technological information provided by academics so that the producers can understand and apply it, to rescue local knowledge from producers and to know participatory and teamwork techniques that facilitate communication and maintain the attention and interest of all group members.

5.2.2 *Training*

Training should be seen as a continuous process of human resources training, both to improve the skills and attitudes of people involved in the implementation of HPM. Skills refer to acquired knowledge and developed tools, while attitudes

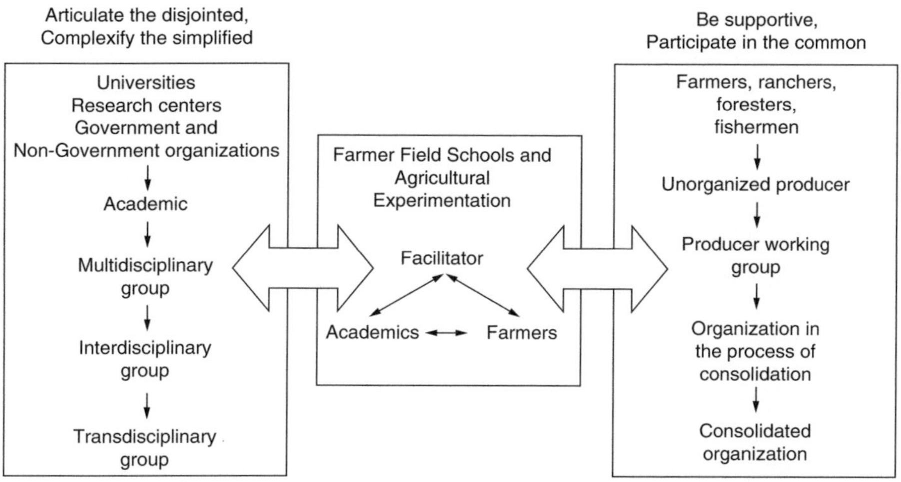

Fig. 5.2 Diagram showing the reorganization of mental system to relearn to learn. In this, the academy is integrated into multi, inter and transdisciplinary groups. On the other hand, producers transit through different degrees of organization, from unorganized producers to consolidated organizations. Both groups – academics and producers – interact through Farmer Field Schools and Agricultural Experimentation, where producers and academics enhance their work by sharing knowledge and experience through facilitators. (Modified from Barrera et al. 2007)

emphasize principles and values. Thus, aptitudes and attitudes form an indispensable binomial for proper functioning of HPM.

According to Morin (2000, 2006), through training, emphasis will be given to reorganizing the mental system *to re-learn to learn*, by articulating the disarticulated and to complexify the simplified, as well as to be supportive and participate in the common. Comprehensive training like this involves the participation of natural science instructors (engineers, biologists, physicians, etc.) and social sciences (sociologists, anthropologists, psychologists, etc.). Multidisciplinary and interdisciplinary are forms of daily work that are promoted from the HPM (Fig. 5.2).

There are several training models among which the Farmer Field Schools and Agricultural Experimentation (FFSAE) model stands out for its affinity with the concept and principles of the HPM. We have been using the FFSAE in trainings processes since the late 1990s. It is also desirable to conclude the training processes by certifying the skills acquired.

5.2.2.1 Farmer Field Schools and Agricultural Experimentation

It is suggested that the training should be carried out under the FFSAE model (Barrera et al. 1999; Braun et al. 2000; Jarquín 2003; Jarquín et al. 2018). This approach combines characteristics of the Farmers Field Schools (FFS) model initially supported by FAO in Indonesia (Williamson 1998; Matteson 2000) that is based on the concept of "*learning-doing*", with the Local Research and

Experimentation Centers (LREC) model developed by the Tropical Agricultural Research Center (CIAT) (Williamson 1999; Braun et al. 2000), which emphasizes agricultural *"experimentation by producers"*.

The "*classrooms*" of the FFSAE are the crop fields or "*farm-schools*". Training topics are offered according to the stages of crop development, so that the group meets several times during a crop cycle. Normally, each training meeting lasts 2 days. On the first day, the producers are trained and on the second day the trained producers replicate the knowledge acquired before a group of producers in the community closest to the farm-school. With this, the producer is trained and at the same time learns to train and share the knowledge acquired.

An important aspect of the process is that, at the end of each training meeting, producers must establish their own FFSAE in their communities to replicate what they have learned and report their progress to the group at the next training meeting. In this way, the training is replicated and knowledge is disseminated with active participation of the producers themselves. Returning to the concept of "IPM community" developed from the FAO FFS in Indonesia and Vietnam (Matteson 2000; Fakih et al. 2003), the FFSAE aim at creating a HPM community with graduates so that, like their counterpart, they collaborate in the dissemination and support with local resources for the HPM implementation.

5.2.2.2 Certification of Acquired Capacities

At the conclusion of the training processes it is suggested to certify the acquired skills. Nowadays, it is widely accepted that the recognition of acquired skills, through the so-called "Labor skills", is a means to increase work performance and motivation (Escobar Valencia 2005; Martínez Clares and Echeverría Samanes 2009). The competences are, according to Mrs. Claude Lévy-Leboyer (cited by Escobar Valencia 2005), "*behavior repertoires that some people dominate better than others, which makes them effective in a given situation*". The concept of competition must be defined more in terms of knowing *how to act*, than knowing *how to do* (IDEA 2000; Le Boterf 2001).

Preferably, it should take advantage of official skills recognition programs such as the National Council for Standardization and Certification of Labor Skills, a parastatal entity sectorized in the Ministry of Public Education of Mexico, which recognizes the knowledge, skills, abilities and attitudes of people, acquired at work or throughout their lives, with national and official certifications (CONOCER 2019).

5.3 The System, Its Limits, Components and Interactions

5.3.1 Brainstorming

According to Andueza (1983), "*brainstorming*" is a technique that consists in producing original ideas through the participation of members of an informal group that, deprived of inhibitions, express themselves freely on a subject. We have found that this technique helps to identify the problems that affect a socio-environmental system and, based on the problems identified, we have been facilitated to identify the components of that system (Barrera et al. 2004, 2013).

To make the system problem as closely as possible to reality, it is suggested that the members of the group should be the affected farmers and other social actors in the production food chain such as extension agents, service providers and representatives of governmental and non-governmental institutions.

Brainstorming is triggered by asking the group to answer questions such as: *What are the main problems that limit producing the "x" product in a sustainable way?* The answers are written on a blackboard or a spreadsheet[1] whose content is projected on a screen so that everyone can see them. All responses are noted without discussion, only duplications are avoided. If necessary, the group can make the decision to join two or more similar problems into one. Since it is a question of identifying the problems, it is suggested to specify precisely the type of problem in question, to avoid confusion.

5.3.2 Prioritization of Problems

Problems identified by group members should be listed in order of priority. To do this, it is requested that each member of the group rates the problems from a greater to a lesser importance, placing "1" for the most important problem, "2" for the one that follows, and so on. The problems qualification must be performed individually to avoid the influence of the opinion of some people on others. To facilitate this process, it is suggested to print the list of problems and distribute it among the group members.

At the end of the exercise, the individual priority lists are put together to calculate the average points obtained for each problem identified. To facilitate the calculations, it is suggested to capture the information in a spreadsheet. The problems that obtain the averages closest to "1" will be the most important.

[1] For example, a Microsoft Excel™ spreadsheet.

5.3.3 Structural Analysis

The analysis of complex systems requires special tools. One of them is "structural analysis" (Mojica 1991, 2004), a prospective technique that allows to determine interactions among the system elements. In our case, these are inferred from the problems identified through brainstorming and subsequent priority. In order to facilitate the application of structural analysis, it is suggested to analyze only the first 10 problems of the list ordered by priority.

To apply a structural analysis, a matrix is drawn up in a spreadsheet with the 10 most important identified problems. Through this matrix the interaction among problems is determined through two variables, interdependence and mobility:

- Interdependence refers to the degree of subordination of a problem with respect to others; and
- Mobility is the force that has a problem with others.

The interaction among problems is assessed by the group of participants by assigning "0" or "1" to refer to "without" and "with" interaction, respectively. The interaction can be established in a more complex manner if one considers its degree, for example: 0 = no interaction, 1 = low, 2 = regular and 3 = high interaction. After completing the matrix, the total points accumulated per problem per column (interdependence) and row (mobility) are calculated. Finally, the total points accumulated by the problem is transformed into a percentage and with this data the interdependence-mobility graphic is prepared. As an example of a matrix for the case of the main problems associated with coffee system in Chiapas (Mexico), Table 5.2 is presented.

The graph that is generated with the structural analysis is a Cartesian plane with four zones that are defined when crossing the averages of the abscissa (interdependence) with the ordinate (mobility) to obtain Fig. 5.3, and which are called:

- Power zone: Problems of high mobility and low interdependence.
- Conflict zone: Problems of high mobility and high interdependence.
- Autonomous zone: Problems of low mobility and low interdependence.
- Exit zone: Problems of low mobility and high interdependence.

Fig. 5.4 is obtained by plotting the data from Table 5.2 as shown in Figs. 5.3. This figure shows the problems of the coffee system distributed by zone according to the relationship between interdependence and mobility.

From the analysis of Fig. 5.4, with the purpose of proposing a management plan for the socio-environmental coffee system, the following can be said:

- The problems located in the power zone, due to their great influence on the others, are those that must be solved first (producer disorganization, unfavorable markets, disarticulated industrialization).
- The problems in the conflict zone exert great influence on other problems, but are highly vulnerable to them (training and technical assistance).

Table 5.2 Matrix of the structural analysis for the problems identified in the coffee growing region of Chiapas, Mexico (modified from Barrera et al. 2004)

		Problems (0 = without interaction, 1 = with interaction)[b]										Mobility	
Num	Problems[a]	1	2	3	4	5	6	7	8	9	10	Total	%
1	Low prices		1	0	1	1	0	1	1	1	0	6	9.8
2	Insufficient economic resources	0		0	1	1	0	1	1	1	0	5	8.2
3	Government support out of time	0	1		1	1	0	1	1	0	0	5	8.2
4	Pests and diseases of coffee	1	1	0		0	1	0	0	0	0	3	4.9
5	Poor technical assistance	0	0	0	1		1	1	1	1	1	6	9.8
6	Unfavorable markets	1	1	0	1	1		1	1	1	1	8	13.1
7	Poor training	1	1	0	1	1	1		1	1	1	8	13.1
8	Poorly grow coffee plantations	1	1	0	1	0	1	0		0	0	4	6.6
9	Producer disorganization	1	1	1	1	1	1	1	1		1	9	14.8
10	Disarticulated industrialization	1	1	0	0	1	1	1	1	1		7	11.5
	Interdependence												
	Total	6	8	1	8	7	6	7	8	6	4	61	100.0
	%	9.8	13.1	1.6	13.1	11.5	9.8	11.5	13.1	9.8	6.6	100.0	

[a]Problems in order of priority

[b]1 Low prices, 2 Insufficient economic resources, 3 Government supports out of time, 4 Pests and diseases that affect coffee, 5 Poor technical assistance, 6 Unfavorable markets, 7 Poor training, 8 Poorly grown coffee plantations, 9 Producer disorganization, 10 Disarticulated industrialization

- The problems that are largely resolved by solving other problems (insufficient economic resources, poorly cultivated coffee and coffee pests and diseases) are located in the exit zone.
- In the area of autonomous problems there are the most independent problems of the system (insufficient and out-of-time Government support and the low international price of coffee).

According to this analysis, the problems caused by pests, located in the exit zone, are characterized by being:

(i) Very interdependent, meaning that, to solve a problem first, the solution of other problems is required; and
(ii) Very low mobility problems, which is interpreted as problems whose solution have a low impact on the solution of other problems of the socio-environmental system.

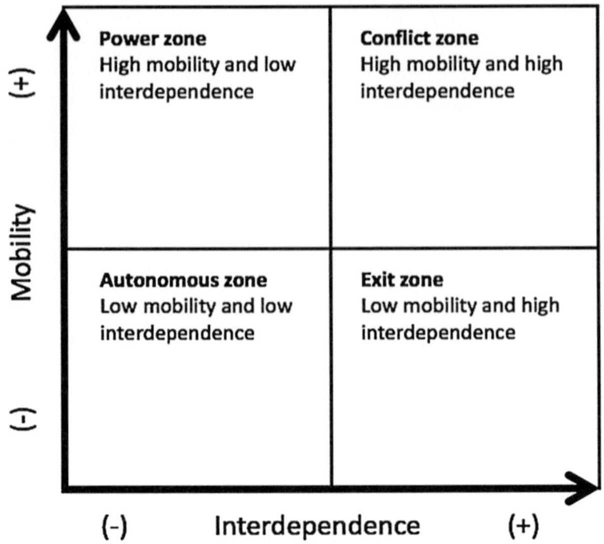

Fig. 5.3 Graph showing the four interdependence-mobility interaction zones that are generated as the final result of the structural analysis. (Modified from Mojica 1991, 2004)

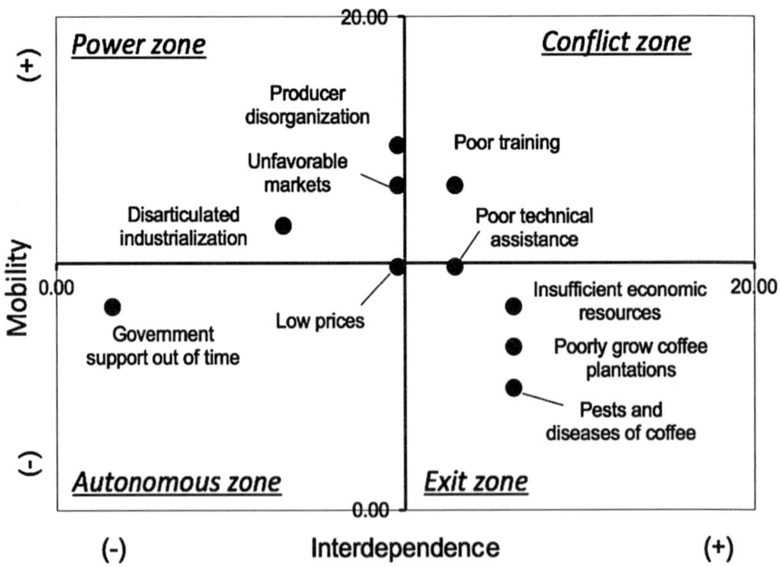

Fig. 5.4 Structural analysis of the 10 most important problems of the socio-environmental systems of coffee growing region of Chiapas. (Modified from Barrera et al. 2004). © 2020 From Holistic pest management by Juan F. Barrera. Reproduced by permission of Taylor and Francis Group, LLC

5.4 Work Plan

Based on the results of the structural analysis, the work team proceeds to prepare the work plan. First, the problems identified and analyzed through structural analysis become the elements or components of the system (Table 5.3). Then, based on Fig. 5.5 that relates the areas of the structural analysis with objectives and strategies, the work plan is prepared. Following this procedure, Fig. 5.6 is obtained, as for the case of the socio-environmental system of coffee in Chiapas. As a last step, the goals, actions, managers, compliance times and their costs for each strategy must be defined (Gamarra Morales 2005).

5.4.1 General Objective

The general objective of the work plan derives from the most felt need, as expressed by farmers. In general, improving incomes is the most perceived need, especially among small producers. Therefore, the general objective is almost always focused

Table 5.3 Conversion of problems to elements and of these to objectives and strategies of the work plan for the management of the socio-environmental systems of coffee growing region of Chiapas, Mexico

Num.	Structural analysis zone	Problems of the system		Elements of the system		Objectives/ strategies
1	Autonomous	Low prices	→	Coffee prices	→	Specific objective
2	Exit	Insufficient economic resources	→	Economic resources (income)	→	General objective
3	Autonomous	Government support out of time	→	Government supports	→	Specific objective
4	Exit	Pests and diseases of coffee	→	Coffee pests	→	Specific objective
5	Exit	Poor technical assistance	→	Technical assistance	→	Specific objective
6	Power	Unfavorable markets	→	Markets	→	Strategy
7	Conflict	Poor training	→	Training	→	Strategy
8	Exit	Poorly grow coffee plantations	→	Coffee plantations	→	Specific objective
9	Power	Producer disorganization	→	Producer organization	→	Strategy
10	Power	Disarticulated industrialization	→	Coffee industrialization	→	Strategy

Prepared from Barrera et al. (2004)

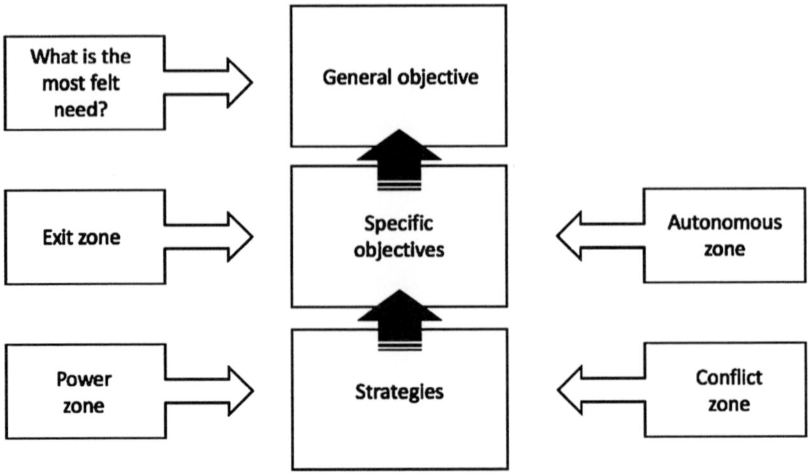

Fig. 5.5 Relationship between the zones of the structural analysis and the objectives and strategies of a work plan for the management of a socio-environmental system

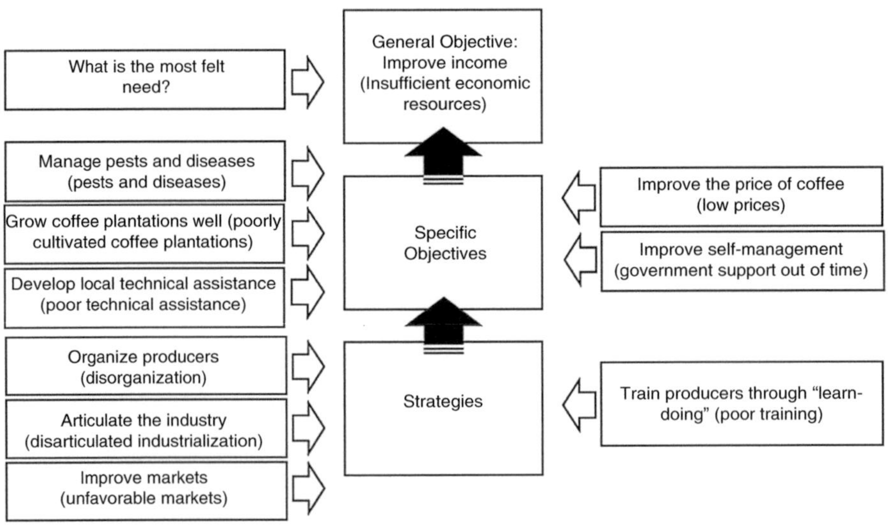

Fig. 5.6 Objectives and strategies of the work plan for the management of the socio-environmental systems of coffee growing region of Chiapas, Mexico. (Prepared from Barrera et al. 2004)

on economic aspects such as improving income. If the overall objective of the work plan does not aim to achieve this need, it will be very difficult for low-income producers to be committed to the work plan.

Although improving the income of producers is usually the work plan priority, in the case of many large producers, especially exporters, the priority is to produce safe products to meet the requirements of the international market. In this case, the objective is usually to avoid or reduce the pollution generated by the production system, both by pathogenic microorganisms (e.g., *Escherichia coli* pathogenic strains) and by excessive use of agrochemicals (maximum residue limits).

For other producers, whose economic (e.g., income) and environmental (e.g., pollution) problems are not limiting, their weakness may be in complying with their obligations to workers (e.g., labor benefits established in the Federal Labor Law) and with the communities surrounding their productive systems (e.g., social responsibility required by product certifiers for special markets). For them, fulfilling social obligations could be the general objective of their work plan. In the case of the richest producers, the challenge does not come from scarcity or need, but from excess production and consumption (Scoones 2017).

5.4.2 Specific Objectives

The specific objectives are elaborated from the problems/elements of the system located in the exit and autonomous zones of the structural analysis. This is done because these problems/elements depend a lot on others. Those located in the exit zone such as pests are, in particular, the most dependent.

5.4.3 Strategies

The strategies are derived from the problems/elements of the system that proceed from the power and conflict zones of the structural analysis. Those problems/elements of the power zone, as the name implies, have the power to move much the system towards the desired condition, without depending so much on the other ones.

5.5 Decision-Making and Monitoring Systems

One of the most important differences between IPM and HPM is the decision-making system. As previously stated, while IPM uses action thresholds such as the Economic Injury Level (EIL) (Stern et al. 1959; Higley and Pedigo 1996), HPM uses the concept of risk through the Holistic Risk Index (HRI) (Barrera et al. 2007; Barrera 2020).

EIL[2] refers to the lowest population density of a pest that could cause economic injury. In other words, it is the pest density at which the cost of the activities carried out for its management, and the benefit generated by the crop, are equal. Using EIL it is possible to determine whether an organism is a pest of economic importance in a given time. Hence, it is considered that without the proper EIL estimate, it is very difficult to avoid economic injury from pests and unjustified insecticide treatments. Thus, EIL is considered the cornerstone of IPM in decision-making.

On the other hand, decisions in HPM are taken from information about the risk[3] coming from both the pests and other elements that make up the system and that interact with the pests. This risk-based systemic gear implies that, under the HPM, decision-making for pest management is based on the organization, structure and operation of the system, that involves many more aspects than the single pests population density and other crop-related variables.

5.5.1 Decision Making System

5.5.1.1 Risk Concept

Before referring to the Holistic Risk Index (HRI), we will see some aspects related to the term "*risk*". First, it must be said that, as a tool for threat prevention and management, risk received a great development in the field of disasters (ITDG 2002; EIRD/ONU 2004; Giupponi et al. 2015). The "disaster risk" is a concept that allows to encompass and summarize the state of a system, with special reference to the human population. It is recognized that disasters hit approximately 75% of the world's population at least once between 1980 and 2000, causing significant deaths and economic losses, especially in developing countries. Because disasters seriously threaten the fulfillment of the Millennium Development Goals, various United Nations programs have carried out very comprehensive studies to design planning tools that help reduce disaster risks (EIRD/ONU 2004; PNUD 2004).

There are many definitions of risk and there are many terms associated with disasters such as earthquakes, hurricanes, floods, droughts or pests (Table 5.4). One of them defines risk as the probability that people may suffer damage due to a disaster (ITDG 2002; Barrera et al. 2007). Another definition of risk is the probability of an adverse event and the magnitude of its consequences (Griffin 2012). In the context of HPM, we will refer to risk as the possibility of a pest happening and its consequences (Barrera 2020).

[2] More information on action thresholds such as the EIL can be found in Chap. 2, Sect. 2.5.3.

[3] More information on risk as a cornerstone of the HPM is discussed in Chap. 4, Sect. 4.6.

Table 5.4 Some definitions of terms associated with disasters

Threat	Probability of occurrence of a potentially destructive or harmful phenomenon or event (earthquakes, floods, landslides, tsunamis, droughts, wars, epidemics) that acts on certain conditions of vulnerability.
Vulnerability	Characteristics of people, groups or societies that determine their degree of exposure to a threat or danger. Key features include class, caste, ethnicity, gender, disability, age and status. Vulnerability is determined by structural causes, social processes and unsafe conditions that interact with each other.
Capacity	Attributes and mechanisms of people, groups, institutions and societies to reduce disaster risks and to survive, resist and recover from the damage caused by such disasters.
Risk	The probability that people may suffer damage due to a disaster.
Disaster	Serious human, material or environmental losses, caused by the impact of a destructive phenomenon on certain conditions of vulnerability.
Preparation	Set of actions and measures to minimize the loss of human lives, and personal and material damage caused by an emergency, organizing timely and effective response and rehabilitation
Emergency	Temporary crisis situation caused by a disaster that puts the lives of people and their assets at serious risk by limiting or preventing basic activities and services. Emergencies are situations that exceed the capacity of the affected society and the institutions to deal with it using only its resources.
Rehabilitation	Implementation of measures to mitigate the damage caused by a disaster through the immediate commissioning of the services and infrastructure of the affected community or society. It involves provisional measures.
Reconstruction	Set of actions aimed at solving the problems generated by disasters. Reconstruction should involve reducing the risk conditions that caused the disasters.
Prevention	Any institutional or citizen action that is undertaken to reduce the risks of a specific population in the face of disasters.
Risk management	Capacity of societies and their social actors to transform risk, acting on the causes that produce it. It includes the measures and forms of intervention that tend to reduce and mitigate or prevent disasters. It also includes disaster management. It is a planned, concerted, participatory and comprehensive process of reducing the disaster risk conditions of a community, a region or a country, linked to the search for sustainability.

Modified from ITDG (2002) and Barrera et al. (2007)

5.5.1.2 Holistic Risk Index

Several ways of estimating risk have been proposed. Based on the work of the ITDG (2002), we propose the Holistic Risk Index (HRI) as a pest risk estimator. The HRI is estimated with information from the threat (T), vulnerability (V) and response capacity (C) using the following conceptual equation (Barrera et al. 2007; Barrera 2020):

$$\text{HRI} = \left(\text{T} + \text{V}\right) / \text{C}$$

This equation is based on the idea that pests constitute a threat to farmers and that they present different degrees of vulnerability to their damages and different degrees of response and recovery capacity to avoid or reduce them. Thus, a farmer is at greater risk towards a pest as the threat and vulnerability are greater and his response and recovery capacity is lower.

Unlike other ways of estimating risk, in which the response capacity is implicit in vulnerability, we have considered it convenient to break it down in order to have greater control over the indicators that define it. This approach to risk estimation allows us to identify C as the response and recovery capacity and T + V as the *non*-response and *non*-recovery capacity, that is, as the inability to respond. Interpreting the HRI in this way has great significance in relation to the socio-environmental system management and its implications in pest management. Hence, HRI is expressed in units that come from the relationship of the inability to respond to the response capacity. Thus, the higher the value of C, the lower will be HRI and vice versa. In view of the above, the HRI is defined as:

> The resulting value between the inability to the ability to respond to threats that, under certain conditions of vulnerability, may affect the well-being of socio-environmental systems.

5.5.1.3 Indicators

Table 5.5 presents the concepts of HRI and its components with examples in the case of threats such as the coffee rust leaf (*Hemileia vastatrix*) and the coffee berry borer (*Hypothenemus hampei*), important coffee pests.

With respect to the threat, it should be noted that in addition to the "Infestation of the farm", the "Infestation of neighboring farms" is considered. This variable is interesting because it allows analyzing the interaction between neighboring farms that, in the case of pests such as rust and borer, is of vital importance due to their high dispersal capacity.

About vulnerability, its characteristics can be divided into two groups of factors. The first refers to personal factors of the producer (e.g., sex, age, schooling) and his household (e.g., family members, income/expenses). The second refers to operational factors related to the production system (e.g., altitude, soil fertility, production costs). It has been found that these types of factors affect the acceptance of technology by farmers and, consequently, their income, food, housing and children education (Regalado López et al. 1996). The sum of both conditions allows defining the degree of exposure of the producers and farms to the mentioned pests.

The response capacity, unlike the vulnerability, that refers to the degree of exposure or status of the producers and their farms to these pests, includes the variables and indicators related to the operation of the socio-environmental system that reduce threat and vulnerability (e.g., organizational capacity, ability to sell coffee in a market with better prices, ability to invest in pest control). The following example shows how to differentiate vulnerability from response capacity: the status of a producer's income at a given time is the degree of vulnerability, while the actions that the producer can take to increase his income at a time given are the response capacity.

Table 5.6 shows some examples of indicators for the socio-environmental system of coffee regions following the nomenclature proposed by de Camino and Müller

Table 5.5 Concepts and examples of the Holistic Risk Index and its components, in the case of coffee[a] pests[b]

Components	Concept	Examples	
Holistic Risk Index	The resulting value among the inability to respond to the ability to respond to threats that, under certain conditions of vulnerability, may affect the well-being of socio-environmental systems.	Holistic Risk Index to coffee berry borer Holistic Risk Index to coffee leaf rust	
Threat	It is the occurrence of a pest that acts on certain conditions of vulnerability.	Farm infestation Neighborhood farm infestation	
Vulnerability	These are the characteristics of the producers and farms, which determine their degree of exposure and, therefore, the state of vulnerability to a pest. Vulnerability is determined by structural causes, social processes and unsafe conditions that interact with each other.	Producer characteristics: Sex Age Health Scholarship Aptitudes Attitudes Perceptions Family group Home Income/ Expenditures	Farm characteristics: Altitude (masl) Terrain slope Sun exposure Soil fertility Cultivated coffee species Production system Coffee yield Coffee quality Production costs
Response capacity	They are the attributes and operational mechanisms of producers, organizations, institutions and society to reduce pest risks and to survive, resist and recover from the damage caused by these pests.	Organization Self-management Participation in collaboration groups and networks Knowledge/information Farm administration Training Technical assistance Workforce Access to timely and sufficient resources Agroecological pest management Diversification of products and markets Value added Intermediaries Environmental and social responsability	

[a]Made with information from Barrera et al. (2007, 2011, 2018)
[b]"Pest" refers to any organism that harms the human being

(1993) and their relationship with the components of HRI. The percentage of leaves with rust (incidence) and the degree of infection (severity) would be indicators related to the threat. Examples of indicators related to vulnerability would be: the level of formal studies of the producer (education), the meters above sea level at

Table 5.6 Example of a set of indicators in the "coffee region" farmland system (based on de Camino and Müller 1993) and its relationship with the HRI components

Category[a]	Element[b]	Descriptor[c]	Indicator[d]	Variables that are related
Endogenous and exogenous resources	Biodiversity (phytopathogenic fungi)	Incidence of rust (*Hemileia vastatrix*) (threat)	Percent of coffee leaves with rust	Number of coffee leaves, presence/absence of rust
Endogenous and exogenous resources	Biodiversity (phytopathogenic fungi)	Severity of rust (*H. vastatrix*) (threat)	Degree of infection	Number of rust pustules per coffee leaf
Endogenous resources	Biodiversity (phytopathogenic fungi)	Rust multiplication (*H. vastatrix*) (vulnerability)	Rate of increase	Increase in the incidence/severity of rust between the beginning and the end of the rust cycle
Endogenous and exogenous resources	Biodiversity (insect pests)	Incidence of borer (*H. hampei*) (threat)	Percentage of coffee berries with borer	Number of coffee berries, presence/absence of borer
Endogenous resources	Age of producer	Years completed by the owner of the farm (vulnerability)	Years of life	Age with knowledge, perceptions, experience, health, etc.
Endogenous resources	Schooling of producer	Formal education of the farm owner (vulnerability)	Level of formal studies	Education with knowledge, perceptions
Endogenous resources	Altitude	Habitat (vulnerability)	Meters above sea level	Altitude and rust
Endogenous resources	Soil	Availability (vulnerability)	Size of coffee plantation (hectares, square meters)	Number of hectares cultivated with coffee, farm size
System operation	Socio-economic management (income)	Net income from coffee sales (response capacity)	Income per crop cycle	Coffee yield, coffee price, production cost
System operation	Socio-economic performance (participation in organizations)	Participation in organic coffee certification program (response capacity)	Participate/not participate	Coffee price, support of the organization (advice, supplies)
System operation	Technical management (inputs)	Use of fungicides against rust (response capacity)	Use/not use	Fungicide use, rust control
System operation	Technical management (intensity of pest/disease management)	Pest/disease management strategy (response capacity)	Technological level of the producer	Management strategy, pest/disease control

[a]A category is an aspect of the system
[b]An element is a part of a category
[c]A descriptor is a significant feature of an element
[d]An indicator is a measure of the effect of the system operation on the descriptor

which the coffee plantation is located (altitude) and the size or area of the coffee plantation (soil availability). Among the indicators of response capacity we can mention: the income per crop cycle, whether it participates in a coffee quality certification program and, whether or not the chemical control of rust is applied.

In order to incorporate the indicators that integrate each of the components of the HRI into the equation HRI = (T + V)/C, as is common in this process (e.g., Griffin 2012), they go through the following stages:

(i) Definition of indicators and data collection. In the case of the HPM, the indicators that integrate the HRI components are defined according to the participatory diagnosis and the work plan. Once the indicators for each component are defined, the data for each producer must be obtained. These can be obtained by interviewing the producers and sampling their productive systems, or by resorting to the information that the farmers possess or to the databases of producer organizations. In the event that real data are not available for some variables, estimated data can be used.
(ii) Determination of the relative importance of the indicators. Considering that some indicators may be more important than others, it is desirable to determine the degree to which each of them contributes to explaining the observed variability of the system. For this, it is recommended to use some multivariate method such as a factorial analysis. If the means to perform this type of statistical analysis are not available, the weights or importance of the indicators can be assumed or determined through focus groups. In the latter case, it can be assumed that the indicators have the same weight.
(iii) Relationship of indicators with threat. Establishing the relationship between the indicators and the threat is a fundamental aspect because it allows to determine the vulnerability or degree of exposure of the producer and its production system to the pest. This requires knowing the behavior of the pest under various circumstances. Preferably, these relationships should be based on information from scientific research or empirical knowledge. In the latter case, hypothetical relationships can be established.
(iv) Standardization of the indicators. Based on the relationships established between the indicators and the pest, the original indicators data are standardized (e.g., converting the original data on a scale of 1–100).

5.5.1.4 Calculation of the HRI and Its Components

The standardized indicators in the previous section are used to calculate the components of HRI[4]: threat, vulnerability and response capacity. At this time, the indicator values are multiplied by the constants generated by the multivariate analysis that determine their weight or importance in explaining the system variability.

[4]An example is given in Chap. 6.

Next, the values obtained from the threat, vulnerability and response capacity are transformed into percentages. If the sum of these percentages is equal to 100 (that means the calculations are correct), the HRI for each producer is calculated using equation

$$\mathrm{HRI} = (\mathrm{T} + \mathrm{V}) / \mathrm{C}.$$

5.5.1.5 HRI Graphic Analysis

The HRI can be represented graphically in several ways; here we will refer to the following three[5]:

(i) Holistic risk triangle. It relates the threat to vulnerability and response capacity through a triangular graph (equilateral triangle). This means that each point on the graph results from the combination of the three variables, whose sum is 100%. Ternary graphs are useful for identifying different regions or phases in the graph (phase diagrams). These graphs have broad applications in several fields, such as glass compositions, refractories, aluminum alloys, stainless steels, welding metallurgy or game theory (Selvaduray 2004; Ponsen et al. 2009). The holistic risk triangle is divided into eight regions that indicate the level of risk on a scale ranging from very low (<1.0) and low (1.0–1.5) to medium (1.51–4.0) and high or very high (> 4.0). Likewise, these regions are classified according to the component with the greatest influence (threat or vulnerability) (Fig. 5.7). Holistic risk triangle data can be statistically analyzed by calculating confidence intervals (e.g., 50%, 90% and 95%) and non-parametric estimation of the probability density function (Cornell 2011; Lawson and Willden 2016; Hamilton and Ferry 2018; Barrera and Valle 2019). The confidence intervals of the values are statistical measures of certainty that are used to infer the risk to the threat that the neighboring producers may not have considered in the study. For its part, the probability density function provides information on the probability that the random variable (T-V-C ratio) falls into a specific region of the space of possibilities, provided by the triangular graph.

(ii) Radial graphics. These graphs are used to show the relative importance of the indicators that make up each component of the HRI. That is, this type of graphs allow identifying the indicators that contribute to lowering or increasing the HRI value.

(iii) Maps. With access to geographic coordinates of productive systems, maps with the values of HRI and its components (threat, vulnerability and response capacity) can be drawn up. And, if the sample size is large enough, through an interpolation method of a Geographic Information System, such as Inverse Distance Weighting (IDW), the values of HRI and its components can be interpolated to other producers with productive systems located in the same geographical region.

[5] More information on the graphic representation of the HRI is presented in Chap. 6.

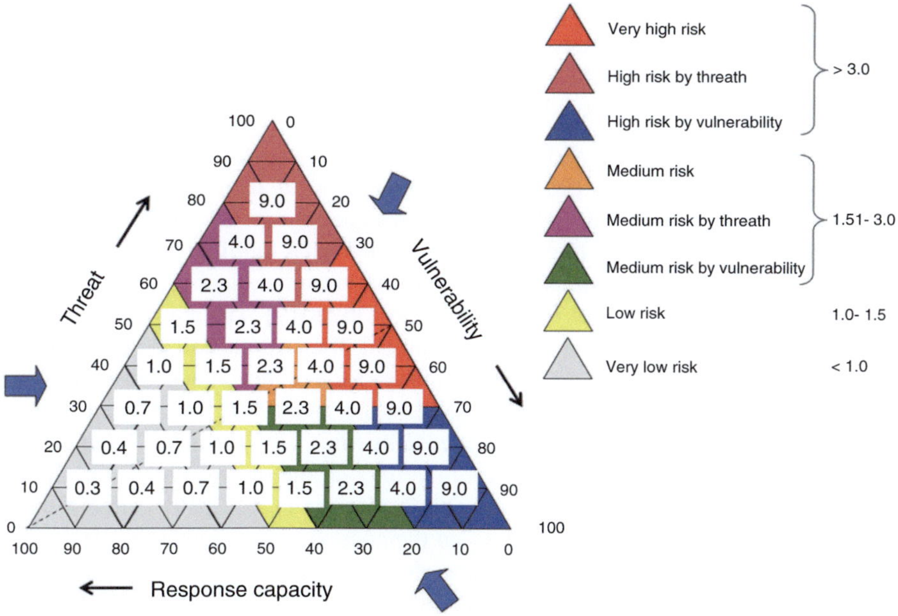

Fig. 5.7 The holistic risk triangle relates the components of the Holistic Risk Index (HRI) using a triangular graph. The numbers inside the triangle are the HRI values calculated with IHR = (T + V)/C, where T = threat, V = vulnerability and C = response capacity. The colors show the areas of the triangular graph with different degrees of risk; dark colors indicate areas with medium and high risks, while lighter colors indicate low and very low risks. The line from the lower left corner of the triangle (union of sides T-C) to the point T = 50%, V = 50% and C = 0%, divides the triangle into two parts both with the same series of values; in the upper part the risks are grouped where T is more important than V, while in the lower part the risks are grouped, where V is more important than T

5.5.2 Monitoring System

Once the indicators have been identified, the first data collection should be carried out, which will give rise to the baseline or current situation of the socio-environmental system. From now on, the work plan of management actions will aim at improving the baseline indicators or, in other words, decrease the HRI by decreasing the threat and vulnerability, while increasing the response capacity. To this end, monitoring of the indicators is essential.

5.5.2.1 Rapid Agroecological Sampling

Rapid Agroecological Sampling (RAS) is a sampling method that consists of collecting socio-economic information through surveys of farmers (and other social actors) and agroecological information through a relatively rapid multidisciplinary samplings. The information collected is used to estimate the HRI and its components. Originally, RAS was developed for the cultivation of coffee, so it must be adapted if it has to be applied to other socio-environmental systems. RAS is a combination of methods that proceed from the following four fields of work:

(i) Sustainable rural livelihoods and Lifeways.
According to Scoones (2017), a Sustainable rural livelihood encompasses capabilities, assets (material and social resources) and activities for a living. While sustainable rural livelihoods emphasize what families have to live each day –capitals and capacities– Lifeways include capital, decision-making strategy, expected results and values (Herrera et al. 2019). The approach of livelihoods and lifeways in the rural environment, focused on holistic and integral perspectives to understand what people do for a living, has been present for decades. However, its boom flourished in the late 1980s. In order to answer questions such as *who owns what, who does what, who gets what, what are the values*, among others, the analysis of livelihoods and lifeways relies on various methods to get answers (Scoones 2017; Herrera et al. 2019). Among these methods and tools, and depending on the context, there are surveys and mappings of various kinds that allow knowledge of social aspects, types of activities, ownership of assets, income and expenses, etc. RAS uses some of these methodologies to gather information about the producer, his home and his production system.

(ii) Environmental services
When properly managed, the ecosystems produce many benefits that make the life of the human being possible on Earth. Such benefits, called environmental services, can be material goods such as wood and seafood, life support processes such as pollination and water purification, and conditions that satisfy life as beauty and serenity. However, the importance of the services provided by ecosystems is only appreciated when they are lost (Daily et al. 2000). To conserve ecosystems, it is necessary to assess and monitor the environmental services they provide. For this, methods have been developed that allow categorizing and quantifying the current state of environmental services, their demand, the relationship with land uses, the effects of agricultural and forestry practices, among others. Medina Fernández et al. (2006) developed four quick and simple methods to assess the status of the following four environmental services in coffee plantations: water conservation, biodiversity, carbon fixation and soil conservation. Depending on the context, some of these methodologies are used in RAS.

(iii) Crop yield estimation
Knowing in advance the performance of the productive system is of great importance for producers and others as agronomists, economists and decision makers. The yield forecast provides relevant information to know the expected income, to plan crop and harvest management and product marketing (Machado 1955; Galán et al. 2008). Also, the yield forecast is an important tool for optimizing crop yields and for evaluating crop area insurance contracts (Basso

et al. 2013). Likewise, the purpose of yield forecast must be to reduce the risks associated with food systems (Bernardi et al. 2016). In the case of coffee, almost any producing country has developed its own methodologies to estimate the yield. Among these, we have used the one developed by the Mexican coffee institution in the 1980s (INMECAFÉ 1985). The method consists of categorizing the plants by age and stage of production and determining the frequency of these categories per unit area. Subsequently, through sampling of branches, the number of fruits per plant in each category and their respective conversion to kg of coffee beans is estimated. Finally, to estimate the expected yield per ha, the estimated production per category is multiplied by the number of coffee plants per ha, in each category.

(iv) Pest management
Since pests represent the "threat" in HPM, methods are required to estimate their incidence and severity on the productive system; in addition to phytophagous arthropods and pathogens and plants (weeds) that compete with crops, it is important to sample the populations of natural enemies of such organisms, as well as beneficial plants. One method that allows obtaining this information on several coffee arthropods and pathogens at the same time is the "integral pest count" developed by the Tropical Agronomic Center for Research and Education (CATIE) (Guharay et al. 2000). This method consists in sampling the presence of insects and phytophagous microorganisms and most common herbaceous plants in this productive system. In the case of coffee crop, phytophagous (arthropods and pathogens) are sampled on leaves, fruits, branches and plant stems. Weeds are sampled using the "shoe tip" method, whereby groups of competing and non-competing plants with coffee trees are sampled, rather than particular species. Unless the symptoms so indicate, in the case of coffee, the roots are sampled for the presence of borers (Coleoptera: Cerambycidae), mealybugs (Homoptera: Pseudococcidae), nematodes, etc. Field sampling is complemented with information from interviews with producers about pest management (Jiménez et al. 2000).

As examples, but not recipes, Table 5.7 raises the key questions and methods to get answers, when applying RAS.

5.6 Risk and Resilience

The ultimate goal of HPM is to increase the resilience of socio-environmental systems by reducing the risk of threatening pests. Resilience is a concept of multiple nuances in its definition, depending on the context in which it is used. For example, Dr. Boris Cyrulnik from clinical ethology defines resilience as "*the resumption of a development after a physical agony*" (Cyrulnik 2011). From agroecology, Dr. Clara Inés Nicholls Estrada provides the following definition of resilience: "*It is the ability of a system to maintain its organizational structure and productivity after a*

Table 5.7 Key questions and some methods used for answers, when the Rapid Agroecological Sampling is applied

Key question	Work field	Methods	References
Who owns what?	Sustainable rural livelihoods	Surveys and social maps; classification of wealth and assets	Scoones (2017)
Who does what?	Sustainable rural livelihoods	Activity mapping; calendar of agricultural and migratory stations	Scoones (2017)
Who gets what?	Sustainable rural livelihoods	Asset ownership surveys; conflict analysis	Scoones (2017)
What do they do with it?	Sustainable rural livelihoods	Surveys of income and expenses; life story	Scoones (2017)
How do groups interact?	Sustainable rural livelihoods	Interconnection analysis; institutional analysis; organizational mapping; gender analysis	Scoones (2017)
What are the capitals?	Sustainable rural livelihoods Lifeways	Participatory surveys and interviews to determine capitals (human, social, natural, financial and physical)	Scoones (2017) and Herrera et al. (2019)
What is the life strategy?	Lifeways	Participatory surveys and interviews to determine cultivated plants and animals (corn, coffee, livestock) and other activities (migration), as well as social learning and territorial innovation	Herrera et al. (2019)
What are the expected results?	Lifeways	Surveys and participatory interviews to determine land ownership, native and healthy diet, organization and self-management	Herrera et al. (2019)
What are the values?	Lifeways	Surveys and participatory interviews to determine harmony, spirituality, respect, fairness and service	Herrera et al. (2019)
How are they shaped by ecologies?	Sustainable rural livelihoods Lifeways Environmental services	Participatory soil surveys; biodiversity mapping; landscape and field history	Daily et al. (2000), Medina Fernández et al. (2006), Scoones (2017) and Herrera et al. (2019)
Do they conserve the water?	Environmental services	Pollution sampling, and water sediments; sampling of erosion evidence	Medina Fernández et al. (2006)
Do they conserve biodiversity?	Environmental services	Sampling of habitat quality; fertilizer and pesticide use sampling	Medina Fernández et al. (2006)
How much carbon do trees and shrubs fix?	Environmental services	Sampling of the density, diameter of stems and height of trees and shrubs	Medina Fernández et al. (2006)
Do they conserve the soil?	Environmental services	Sampling of soil cover, erosion and soil conservation works	Medina Fernández et al. (2006)

(continued)

Table 5.7 (continued)

Key question	Work field	Methods	References
What is the forecast of the harvest?	Yield estimation	Sampling of the expected harvest	INMECAFÉ (1985)
Which pests predominate?	Pest management	Integral pest count; "shoe tip" weed sampling	Guharay et al. (2000)
What methods do they use for pest control?	Pest management	Producer interview	Jiménez et al. (2000)
Do they receive technical assistance?	Pest management	Producer interview	Jiménez et al. (2000)

disturbance" (Nicholls 2013). The United Nations Development Program (UNDP) defines the construction of resilience as a "*transformative process of strengthening the capacity of men, women, communities, institutions and countries to anticipate, prevent, recover and transform after shocks, tensions and changes*" (Winderl 2014).

Expressions such as:

"… the resumption of a development …"
"… the ability … to maintain …" and,
"… the ability … to anticipate, prevent, recover and transform …"

are reactions to physical agony, a disturbance or shocks, tensions and changes. Interpreting it differently, resilience is largely due to the response capacity of human beings, communities, institutions and countries to factors and phenomena that impact them.

Starting from the above, if we know the response capacity we will be in a position to know the approximate meaning of resilience. This also leads us to establish a relationship between risk and resilience through the Holistic Risk Index (HRI) equation, since the response capacity (C) –together with the threat (T) and vulnerability (V)– is one of the components of risk: IHR = (T + V)/C.

That is, through the HRI equation we can quantify the response capacity and with it, have an approximate measure of resilience. Mathematically, using the HRI equation we can establish the relationship between risk and resilience as follows:

If HRI is the risk (R), then we have:

$$R = (T + V) / C$$

and, if the response capacity C ≈ resilience (E).

Then we replace C with E:

$$R = (T + V) / E$$

Therefore, to estimate the term E, which represents an approximation of resilience, we pass it to the other side of the equation:

$$E = (T + V)/R$$

By plotting R with E, we obtain Fig. 5.8, which shows that there is a power relationship ($y = a.x^b$) between both variables. Assuming that the response capacity is an approximation to resilience, then, by knowing the risk we can estimate an approximate value of resilience (E) with the equation:

$$E \approx 54.325 * R^{-817}$$

where a = 54,325 and b = −817 are constants.

Based on Fig. 5.8, the risk-resilience scale is presented in Table 5.8. As you can see, as the risk falls – following a power ratio ($y = a \cdot x^b$) – the approximate measure of resilience increases and vice versa. In this case, high risks start to appear from values equal to or greater than 3.7, while high values of the approximation of resilience start to appear from 40. Because of their practical importance, other authors have made interesting contributions to understand and measure the risk-resilience relationship in the management of socio-environmental systems (Altieri 2013; Henao-Salazar 2013; Montalba et al. 2013; Gazzano et al. 2015; Machado-Vargas et al. 2018). One of our contributions on the subject is that the risk-resilience relationship is not linear as can be seen in Fig. 5.8.

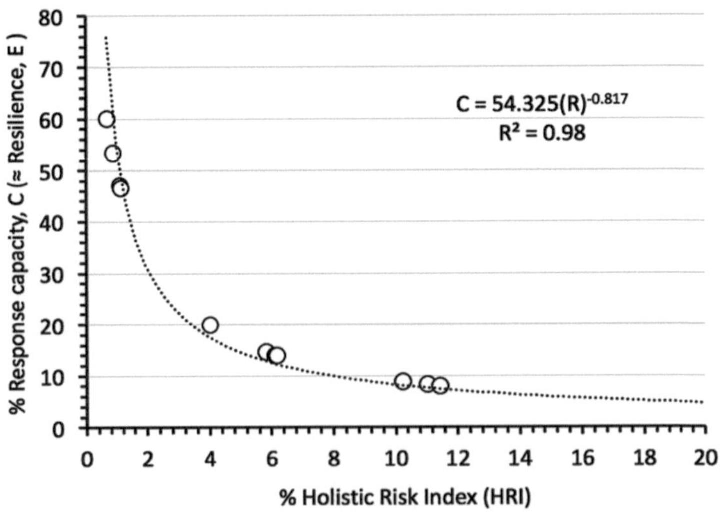

Fig. 5.8 Power relationship ($y = a \cdot x^b$) between risk (R) and response capacity (C) as an approximation of resilience (E) (Barrera 2020). © 2020 From Holistic pest management by Juan F. Barrera. Reproduced by permission of Taylor and Francis Group, LLC

Table 5.8 Scale that relates risk (R) to response capacity as an approximation of resilience (E)

Risk (R)[a]		An approximation of resilience (E)	
Category	Values	Values [b]	Category
Very low	< 1.03	> 50	Very high
Low	1.03–1.4	40–50	High
Medium	1.41–3.6	20–39	Medium
High	3.7–8.5	10–19	Low
Very high	> 10	< 10	Very low

[a]It refers to the Holistic Risk Index (HRI)
[b]The values were defined with $E \approx 54.325 * R^{-817}$ (Fig. 5.8)

5.7 Conclusion

Implementing the HPM requires methods or tools that allow obtaining information on the organization, structure and operation of the socio-environmental system that we want to study or manage. By way of proposals – but not recipes because it will depend on the context – in this chapter some methods help to get answers to the questions posed (Scoones 2017). The methods listed provide socio-economic and environmental information, since it is necessary to know how producers earn a living and how they manage their productive system to deal with pests. Risk-based decision making – Holistic Risk Index – and its relationship with resilience, is one of the most important items in the HPM methodology. Hence, defining the most appropriate indicators that allow monitoring and analyzing risk and its components – threat, vulnerability and response capacity – is of vital importance.

References

Altieri, M. A. (2013). Construyendo resiliencia socio-ecológica en agroecosistemas: Algunas consideraciones conceptuales y metodológicas. In C. I. Nicholls, L. A. Ríos-Osorio, & M. A. Altieri (Eds.), *Agroecología y resiliencia socioecológica: Adaptándose al cambio climático* (pp. 94–104). Medellín- REDAGRES, CYTED y SOCLA. Colombia.

Andueza, M. (1983). *Dinámica de grupos en educación* (3rd ed.). Editorial Trillas SA de CV Mexico.

Barrera, J. F. (2006). Manejo holístico de plagas: Hacia un nuevo paradigma de la protección fitosanitaria. In J. Pohlan, L. Soto, & J. Barrera (Eds.), *El cafetal del futuro: Realidades y Visiones* (pp. 63–82). Aachen: Shaker Verlag.

Barrera, J. F. (2020). Holistic Pest management. In D. Perez-Staples, F. Diaz-Fleischer, P. Montoya, & M. T. Vera (Eds.), *Area-wide management of fruit fly pests* (pp. 275–293). Boca Raton: CRC Press, Taylor & Francis Group.

Barrera, J. F., & Valle, J. (2019). Triángulo holístico de riesgo: Intervalos de confianza y función de densidad de probabilidad. In *XXIV Simposio Latinoamericano de Caficultura.* Guatemala, C.A, p. 460.

Barrera, J. F., Junghans, C., & Jarquín, R. (1999). Programa piloto de escuelas de campo para agricultores: Una propuesta para México. In E. Rodríguez Leyva & E. E. Escobar Aguayo (Eds.), *XXII Congreso Nacional de Control Biológico,* Montecillo, Estado de Mexico, Mexico, pp. 258–261.

Barrera, J. F., Parra Vázquez, M., Herrera Hernández, O. B., et al. (2004). *Plan Estatal de Manejo Agroecológico del Café en Chiapas: Guía hacia una cafeticultura sustentable*. Tapachula: Comisión para el Desarrollo y Fomento del Café de Chiapas and El Colegio de la Frontera Sur, 164 p.

Barrera, J. F., Herrera, J., & Gómez, J. (2007). Riesgo-vulnerabilidad hacia la broca del café bajo un enfoque de manejo holístico. In J. F. Barrera, A. García, V. Domínguez, & C. Luna (Eds.), *La Broca del Café en América Tropical: Hallazgos y Enfoques* (pp. 131–141). Tapachula: Sociedad Mexicana de Entomología and El Colegio de la Frontera Sur.

Barrera, J. F., Gamboa, W., Gómez, J., et al. (2011, August 17–19). Método holístico para la toma de decisiones en el manejo de plagas. In *Simposio Estado del arte del Manejo Ecológico de Plagas en América Latina*, Tercer Congreso Latinoamericano de Agroecología, Oaxtepec, Mexico.

Barrera, J. F., Gómez-Ruiz, J., Parra-Vázquez, M. R., et al. (2013). Análisis de problemas ocasionados por plagas del café bajo un enfoque holístico. *Entomología Mexicana, 12*(2), 1128–1133.

Barrera, J. F., Valle, J., Gómez, J., et al. (2018). *Manejo holístico de plagas en zonas cafetaleras: Concepto y método*. Universidad Autónoma Chapingo- Centro Nacional de Investigación, Innovación y Desarrollo Tecnológico del Café-El Colegio de la Frontera Sur. Mexico, 42 p.

Basso, B., Cammarano, D., & Carfagna, E. (2013). *Review of crop yield forecasting methods and early warning systems* (Report to the global strategy to improve agricultural and rural statistics). Rome: FAO.

Bernardi, M., Delincé, J., Durand, W., et al. (2016). *Crop yield forecasting: Methodological and institutional aspects*. Rome: FAO.

Braun, A. R., Thiele, G., & Fernandez, M. (2000). *Farmer field schools and local agricultural research committees: Complementary platforms for integrated decision-making in sustainable agriculture*. Agricultural Research and Extension Network, ODI. AgREN Network Paper No. 105. London.

CONOCER (Consejo Nacional de Normalización y Certificación de Competencias Laborales). (2019). *¿Qué es el CONOCER? Misión, Visión, Política y Objetivos de calidad.* https://conocer.gob.mx/acciones_programas/conocer-mision-vision-politica-objetivos-calidad/#. Accessed 29 Sept 2019.

Cornell, J. (2011). *A primer on experiments with mixtures*. Hoboken: Wiley.

Cyrulnik, B. (2011). Criterios de resiliencia: Condiciones de un nuevo desarrollo después del traumatismo. In A. G. Sánchez & L. Gutiérrez (Eds.), *Criterios de resiliencia entrevista a Boris Cyrulnik* (pp. 53–72). Barcelona: Gedisa Editorial.

Daily, G. C., Söderqvist, T., Aniyar, S., et al. (2000). The value of nature and the nature of value. *Science, 289*(5478), 395–396.

de Camino, R., & Müller, S. (1993). *Sostenibilidad de la agricultura y los recursos naturales: bases para establecer indicadores*. Instituto Interamericano de Cooperación para la Agricultura/Proyecto IICA/GTZ. Serie Documentos de Programas No. 38. San José, Costa Rica, 134 p.

EIRD/ONU (Secretaría Interinstitucional de la Estrategia Internacional para la Reducción de Desastres, Naciones Unidas). (2004). *Vivir con el Riesgo*. Informe mundial sobre iniciativas para la reducción de desastres, Ginebra, Suiza.

Escobar Valencia, M. (2005). Las competencias laborales: ¿La estrategia laboral para la competitividad de las organizaciones? *Estudios Gerenciales, 96*, 31–55.

Fakih, M., Rahardjo, T., & Pimbert, M. P. (2003). *Community integrated pest management in Indonesia*. Brighton: International Institute for Environment and Development and Institute for Development Studies.

Galán, C., García-Mozo, H., Vázquez, L., et al. (2008). Modeling olive crop yield in Andalucía, Spain. *Agronomy Journal, 100*(1), 98–104.

Gamarra Morales, F. (2005). *Análisis FODA y estrategias para instituciones educativas*. https://es.calameo.com/read/000661059f9e5766ad5d7. Accessed 15 July 2019.

García, R. (1994). Interdisciplinariedad y sistemas complejos. In E. Leff, R. García, P. Gutman, V. M. Toledo, H. M. C. Vessuri, R. Fernández, & R. Brañes (Eds.), *Ciencias Sociales y Formación Ambiental* (pp. 85–124). Mexico: Gedisa Editorial.

Gazzano, I., Altieri, M., Achkar, M., et al. (2015). Holistic risk index: A case study of cattle producers in the protected área of Farropos Estuaries-Uruguay. *Agroecology and Sustainable Food Systems, 39*(2), 209–223.

Giupponi, C., Mojtahed, V., Gain, A. K., et al. (2015). Integrated risk assessment of water related disasters, the SERRA approach. In P. Paron & G. Di Baldassarre (Eds.), *Hydro-meteorological hazards and disasters* (pp. 163–200). Amsterdam: Elsevier.

Griffin, R. (2012). Basic concepts in risk analysis. In C. Devorshak (Ed.), *Plant pest risk analysis: Concepts and application* (pp. 7–18). Wallingford: CAB International.

Guharay, F., Monterrey, J., Moterroso, D., et al. (2000). *Manejo integrado de plagas en el cultivo del café* (Serie Técnica. Manual Técnico No. 44). Managua: Centro Agronómico Tropical de Investigación y Enseñanza-CATIE, 267 p.

Hamilton, N. E., & Ferry, M. (2018). ggtern: Ternary diagrams using ggplot2. *Journal of Statistical Software, Code Snippets, 87*(3), 1–17.

Hart, R. D. (1985). *Conceptos básicos sobre agroecosistemas*. Turrialba: Centro Agronómico Tropical de Investigación y Enseñanza, 159 p.

Henao-Salazar, A. (2013). Propuesta metodológica de medición de la resiliencia agroecológica en sistemas socio-ecológicos: Un estudio de caso en los Andes Colombianos. *Agroecología, 8*(1), 85–91.

Herrera, O. B., Parra, M., Livscovsky, I., et al. (2019). Lifeways and territorial innovation: Values and practices for promoting collective appropriation of territory. *Community Development Journal, 54*(3), 427–445.

Higley, L. G., & Pedigo, L. P. (1996). The EIL concept. In L. G. Higley & L. P. Pedigo (Eds.), *Economic thresholds for integrated pest management* (pp. 9–21). Lincoln: University of Nebraska Press.

Hilje, L., & Ramírez, O. (1992). Una propuesta comprensiva para el desarrollo de programas de Manejo Integrado de Plagas (MIP) en América Central. *Manejo Integrado de Plagas, 24–25*, 63–71.

IDEA. (2000). *La gestión por competencias*. http://www.guyleboterf-conseil.com/IDEA.PDF. Accessed 29 Sept 2019.

INMECAFÉ (Instituto Mexicano del Café). (1985). *Estimación de cosecha de café*. Notes on coffee delivered by the different managements of the Directorate of Coffee Production and Improvement, in the meeting held with the Heads of Department on January 31 and February 1, 1985, Mexico, 13 p.

ITDG (Intermediate Technology Development Group). (2002). *El fenómeno de El Niño y la gestión de riesgo de desastres*. Taller de capacitación. Proyecto Moquegua. Ministerio Alemán- Agro Acción Alemana. ITDG-CEOP-ILO.

Jarquín, R. (2003). Las ECEAs: Base para la implementación de proyectos de desarrollo autogestionarios en zonas cafetaleras. *LEISA Revista de Agroecología, 19*, 33–36.

Jarquín, R., Ramírez Tobías, H. M., Aguilar Benítez, G., et al. (2018). *Las Escuelas de Campo y Experimentación para Agricultores (ECEA's): Guía metodológica y logros alcanzados en Chiapas y San Luis Potosí*. San Luis Potosí: Universidad Autónoma de San Luis Potosí.

Jiménez, L., Barrera, J. F., Rojas, M., et al. (2000). Tácticas y estrategias usadas por ejidatarios para combatir a la "Broca del café" *Hypothenemus hampei* (Coleoptera: Scolytidae), en el muni-

cipio de Tapachula, Chiapas. In *XXIII Congreso Nacional de Control Biológico*, Guanajuato, Mexico, pp. 208–210.
Lawson, J., & Willden, C. (2016). Mixture experiments in R using mixexp. *Journal of Statistical Software, Code Snippets, 72*(2), 1–20.
Le Boterf, G. (2001). *Ingeniería de las competencias*. Barcelona: Ediciones Gestión 2000, S.A.
Machado, A. (1955). Pronóstico de las cosechas del cafeto. Algunos aspectos de los primeros trabajos exploratorios. *Revista Facultad Nacional de Agronomía Medellín, 16*(47), 67–93.
Machado-Vargas, M. M., Nicholls-Estrada, C. I., & Ríos-Osorio, L. A. (2018). Social-ecological resilience of small-scale coffee production in the Porce river basin, Antioquia (Colombia). *Idesia, 36*(3), 141–151.
Martínez Clares, P., & Echeverría Samanes, B. (2009). Formación basada en competencias. *Revista de Investigación Educativa, 27*(1), 125–147.
Matteson, P. C. (2000). Insect pest management in tropical Asian irrigated rice. *Annual Review of Entomology, 45*, 549–574.
Medina Fernández, B. Y., Muñoz Astaíza, C. Y., Haggar, J., et al. (2006). *Metodología para la evaluación de servicios ambientales*. Asociación Nacional del Café. Foreign & Commonwealth Office London, 36 p.
Mojica, S. F. (1991). *La prospectiva; técnicas para visualizar el futuro*. Bogotá: Fondo Editorial Legis.
Mojica, F. J. (2004). *El modelo prospectivo llevado a la práctica*. Primera versión. Convenio Andrés Bello (CAB). Serie: Documentos de Ciencia, Tecnología e Innovación de los países del CAB. Bogotá, Colombia, 178 p.
Montalba, R., García, M., Altieri, M., et al. (2013). Utilización del Índice Holístico de Riesgo (IHR) como medida de resiliencia socioecológica a condiciones de escasez de recursos hídricos. Aplicación en comunidades campesinas e indígenas de la Araucanía, Chile. *Agroecología, 8*(1), 63–70.
Montes, L. M., & Leff, E. (1986). Perspectiva ambiental del desarrollo del conocimiento. In E. Leff (Coord) *Los problemas del conocimiento y la perspectiva ambiental del desarrollo* (pp. 22–44). Mexico: Siglo XXI Editores.
Morin, E. (2000). *La mente bien ordenada: Pensar la reforma, reformar el pensamiento*. Barcelona: Ed. Siex Barral S.A, 185 p.
Morin, E. (2006). *El Método* (7a. ed.). Madrid: Ediciones Cátedra, 448 p.
Nicholls, C. I. (2013). Enfoques agroecológicos para incrementar la resiliencia de los sistemas agrícolas al cambio climático. In C. I. Nicholls, L. A. Ríos Osorio, & M. A. Altieri (Eds.), *Agroecología y resiliencia socioecológica: Adaptándose al cambio climático* (pp. 18–29). Medellín: CYTED.
PNUD (Programa de las Naciones Unidas para el Desarrollo). (2004). *La reducción del riesgo de desastres: Un desafío para el desarrollo*. Informe Mundial. Dirección de Prevención de Crisis y Recuperación, New York, 147 p.
Ponsen, M., Tuyls, K., Kaisers, M., et al. (2009). An evolutionary game-theoretic analysis of poker strategies. *Entertainment Computing, 1*, 39–45.
Regalado López, J., Jiménez Sánchez, L., Laird, R. J., et al. (1996). Factores asociados a la utilización de la tecnología de alta productividad entre productores de maíz de temporal. *Agrociencia, 30*(1), 139–145.
Reichlderfer, K. H. (1980). Economics of integrated pest management: Discussion. *The American Journal of Agricultural Economics, 62*(5), 1012–1013.
Scoones, I. (2017). *Medios de vida sostenibles y desarrollo rural*. Barcelona: Icaria Editorial SA.
Selvaduray, G. (2004). *Ternary phase diagrams: An introduction*. San Jose: San Jose State University.

Stern, V. M., Smith, R. F., van den Bosch, R., et al. (1959). The integrated control concept. *Hilgardia, 29*(2), 81–101.

Williamson, S. (1998). Understanding natural enemies; review of training and information in the practical use of biological control. *Biocontrol News and Information, 19*, 117N–125N.

Williamson, S. (1999). Challenges for farmer participation in coffee research and extension. In E. Rodríguez Leyva & E. E. Escobar Aguayo (Eds.), *XXII Congreso Nacional de Control Biológico*, Montecillo, Estado de Mexico, Mexico, 252 pp.

Winderl, T. (2014). *Disaster resilience measurements: Stocktaking of ongoing efforts in developing systems for measuring resilience*. United Nations Development Programme, 59 p.

Chapter 6
Implementing Holistic Pest Management

> *"Crop protection must be seen within a much broader context, and its location in any specific situation must be determined in collaboration with the local population.*
> *How will a real sociocentric approach to crop protection be implemented?"*
>
> *Stephen Morse and William Buhler, The need for another paradigm shift (Morse S, Buhler W (1997) Integrated pest management: Ideals and realities in developing countries. Lynne Rienner Publishers, London, p. 145, 149).*

Abstract This chapter presents an illustrative and simple guide to implement Holistic Pest Management (MHP). It begins by describing preventive tactics such as the management strategy of pest organisms under the holistic approach. Next, the steps of the critical path of HPM to assess socio-environmental systems are presented. This critical route begins by assessing – with farmer participation – the initial state of the system. It then proceeds through the definition of indicators to calculate the Holistic Risk Index (HRI) – the cornerstone in decision-making of HPM– and concludes evaluating the effectiveness of adaptation and mitigation actions included in the work plan, in order to meet the goals set for system management. Subsequently, the steps for HPM implementation are described. To achieve this goal, emphasis is given to the importance of integrating working groups composed of farmers, academics and facilitators and to the training of the working group through Farmer Field Schools and Agricultural Experimentation. As an example, a case study is presented to estimate and analyze, graphically and statistically, the HRI towards coffee leaf rust.

Keywords Holistic approach · Farmer participation · Holistic Risk Index · Farmer Field Schools · Coffee leaf rust

J. F. Barrera, *Beyond IPM: Introduction to the Theory of Holistic Pest Management*, Sustainability in Plant and Crop Protection 15,
https://doi.org/10.1007/978-3-030-43370-3_6

6.1 Introduction

In the previous chapters the basis for understanding the theory, concept, principles and methods of Holistic Pest Management (HPM) have been provided. We attempted to justify this new paradigm, but also to teach how to think holistically. Now, it is time to integrate what has been previously said to give rise to a coherent and clear guide that helps to implement HPM. In this chapter we aim at teaching how to act holistically.

Like any guide, this contribution is not intended to be prescriptive or exhaustive, but rather being illustrative and simple so that it facilitates identifying and traveling viable paths, for those interested in practicing pest management under a holistic approach.

We will begin by exposing the pest management strategy through HPM. Next, we describe the path to assess the system to study or manage; later, we will see one by one the steps needed to implement HPM. Finally, and by means of one example, we will close the chapter with a case study to illustrate the methodology used to estimate the Holistic Risk Index (HRI), the cornerstone of HPM.

Although the topics that will be discussed later have been the product of our experience with coffee producers (Barrera et al. 2018; Barrera 2020), they can be applied to many other socio-environmental systems, with the corresponding adaptations.

This chapter is closely related to the preceding one (*Methods for the analysis of Holistic Pest Management*), so it is advisable to consult Chap. 5 when more explanations are required on the methods herein used.

6.2 Holistic Management Strategy for Pest Organisms

The pest management strategy under the HPM approach relies on giving a priority to preventive tactics, followed by direct action tactics if preventive measures are not sufficient (Fig. 6.1).

Thus, this strategy promotes prevention rather than cure. To a large extent, preventive tactics must be compatible with natural processes and cycles, in particular, they must preserve and protect natural enemies of pests through soil and vegetation management (Altieri and Trujillo 1987; Andow 1991; Altieri 1993, 1994; Altieri and Nicholls 2003, 2007). Agroecosystems benefit from the conservation and promotion of biodiversity; hence, the alteration of biodiversity, which is responsible for valuable environmental services, can result in the loss of useful species or increase in harmful ones (Chapin et al. 2000). In other words, it is required to produce agricultural products while conserving natural resources (Barrera 2020). This approach is based on the stabilization of production, rather than its maximization and, consequently, the crop can lose productivity. However, it sustains other important aspects of the system such as the capacity for multiple use of the habitat (Altieri 2018).

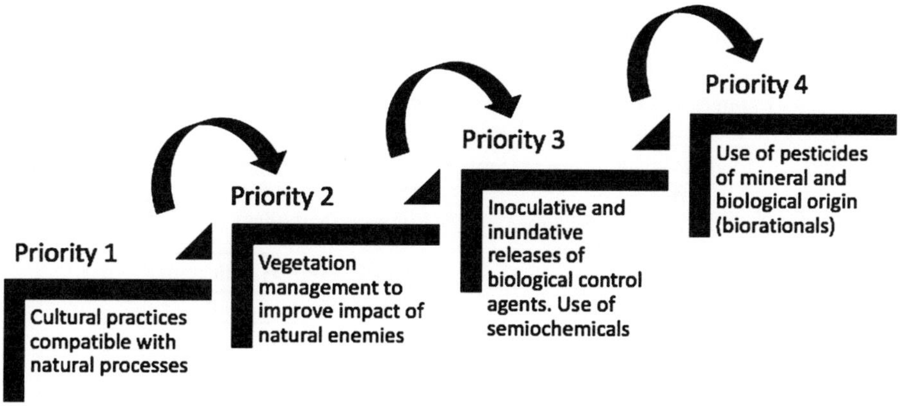

Fig. 6.1 Management strategy for pest organisms under the HPM approach. (Modified from Wyss et al. 2005; Zehnder et al. 2007)

Among the direct control measures, HPM favors the use of biological control agents (parasitoids, predators and pathogens), semiochemicals (e.g., attractants and repellents) and physiological and ecologically selective insecticides (Fig. 6.1). In summary, the HPM strategy is to design systems that, while preventing pest population outbreaks, promote maximum performance and quality with minimum social and environmental damage.

6.3 Critical Path to Assess the System

Figure 6.2 shows the critical path that is suggested to follow in order to correctly assess the system under study or management. It is a cyclic path that should be followed periodically (e.g., based on crop cycles). The path may be partitioned in the following 10 main stages (Barrera et al. 2018):

(i) Evaluate the initial state of the system
(ii) Make a participatory diagnosis
(iii) Identify the problems
(iv) Rate the system
(v) Define the indicators
(vi) Calculate the HRI and its components
(vii) Define the magnitude of risk
(viii) Develop the decision-making process
(ix) Define the work plan and adaptation and mitigation actions
(x) Evaluate the effect of management on the initial state of the system

The purpose of the path is to assess whether the adaptation and mitigation actions defined in the work plan meet the goals of reducing risk and increasing the system's

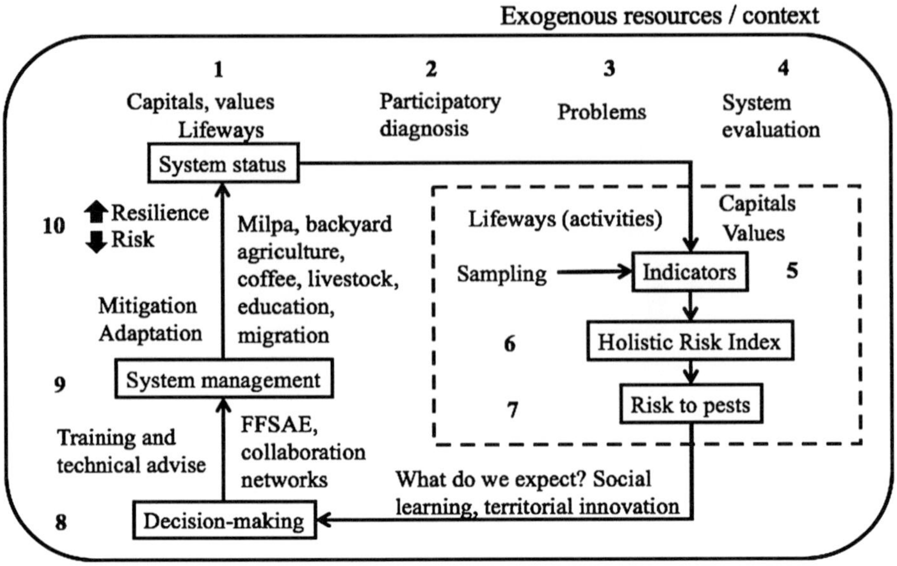

Fig. 6.2 Path to assess the system and decision-making in HPM. (Modified from Barrera 2020) © 2020 From Holistic pest management by Juan F. Barrera. Reproduced by permission of Taylor and Francis Group, LLC

resilience. That is if, on one hand, these actions reduced the threat and vulnerability and, on the other, they increase the response capacity of the socio-environmental system.

6.3.1 Evaluate the Initial State of the System

The evaluation of the initial state of the system allows to know the state of risk and resilience before carrying out any adaptation or mitigation action. The initial evaluation will provide the baseline system assessment.

6.3.2 Make a Participatory Diagnosis

The diagnosis is fundamental to know the problems that afflict the socio-environmental system. For greater success in identifying the main problems, the diagnosis must be made with participation of the people involved in the system, mainly producers.

6.3.3 *Identify the Problems*

Through the identification of the most important problems, the limits of the socio-environmental system are determined. System boundaries allow to concentrate on the most relevant elements (and their interactions) that affect the properties and dynamics of the system. The structural analysis is a very useful tool at this stage (Mojica 1991, 2004).

6.3.4 *Evaluate the System*

The system assessment is carried out using tools such as Rapid Agroecological Sampling (RAS) and lifeways. RAS provides agroecological information on the productive system, while lifeways provide information on capital (human, financial, physical, natural and social), decision-making strategy, expected results and values.

6.3.5 *Define the Indicators*

The variables are derived from the identified problems, and from these, the indicators. The indicators may vary according to the problem or objective of the analysis; however, they must be robust (that meet the conditions for the descriptors that were designed and that have a sufficient measurement basis) and not exhaustive (that refer to the categories and most significant elements of the system) (de Camino and Müller 1993). Once the indicators are defined, their values are obtained by sampling the socio-environmental system, through the RAS and lifeways.

6.3.6 *Calculate the HRI and Its Components*

The RAS and lifeways provide the baseline data to calculate the HRI and its components. In turn, if the risk is known, it is possible to estimate an approximation of resilience.

6.3.7 *Define the Magnitude of Risk*

Through the use of graphic analysis – such as the Holistic Risk Triangle (THR), radial graphs and maps – the socio-environmental system risk magnitude is determined.

6.3.8 Develop the Decision-Making Process

The magnitude of the risk is an essential information for decision-making. The latter, through risk, is the cornerstone of HPM. Questions such as "*What do we expect? What is social learning or territorial innovation?*" help make the right decisions.

6.3.9 Define the Work Plan and Adaptation and Mitigation Actions

The what, who, how, when, how much and where should be defined in the work plan. At this stage, the adaptation and mitigation actions to be taken to reduce the risk and increase the resilience of the socio-environmental system are defined. Mitigation actions are those designed to affect the probability of a result, while adaptation considers actions designed to affect the value of a result (Perrings 2005). Before starting the work plan, it is necessary to train people through, for example, Farmer Field Schools and Agricultural Experimentation (FFSAE), and to develop local technical assistance capabilities. Also, at this stage, collaboration networks are established to facilitate the exchange of knowledge and experiences.

6.3.10 Evaluate the Effect of Management on the Initial State of the System

Once the work plan has been defined and people are trained, adaptation and mitigation actions are initiated. In the case of coffee farmers, these actions aimed at improving the conditions of the milpa,[1] backyard orchards, coffee cultivation and cattle raising, among others, to improve income, education and health. However, they also aimed at reducing forced migration of producers. In order to know if these actions were effective, periodic evaluations of the system should be carried out. If the evaluation shows that the risk fell, according to the programmed goals, it can be concluded that the system increased its resilience to pests.

6.4 Steps to Implement the HPM

Operationally speaking, an HPM program is implemented according to the following five steps (Fig. 6.3):

[1] Milpa means maize field but refers to a field planted not only with maize but with squash, beans, and other plants as well.

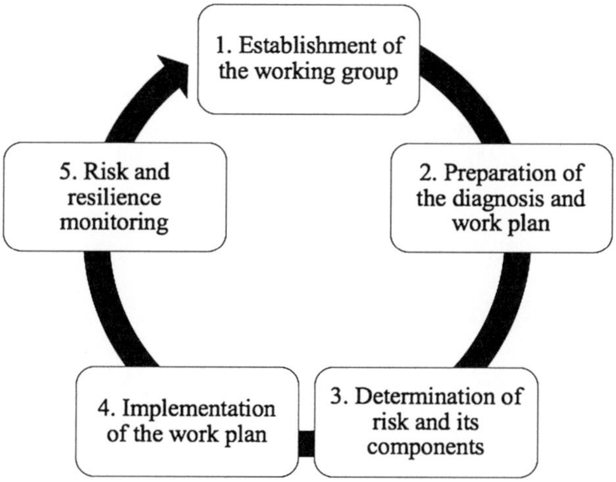

Fig. 6.3 Steps to implement the HPM. (Modified from Barrera et al. 2018)

(i) Establishment of the working group
(ii) Preparation of the diagnosis and work plan
(iii) Determination of risk and its components
(iv) Implementation of the work plan
(v) Risk and resilience monitoring

6.4.1 Establishment of the Working Group

The work team is the basis of the HPM program, as its members will be responsible for carrying out the management and monitoring of the socio-environmental system. Ideally, the team should consist of producers, facilitators and academics.

6.4.2 Preparation of the Diagnosis and Work Plan

The teams make the diagnosis to know the problems of the farmers; with the method of brainstorming, problems are identified and prioritized. Then, the interactions between the problems with the structural analysis are established and, from here, the objectives, strategies and actions of adaptation and mitigation of the work plan are determined.

6.4.3 Determination of Risk and Its Components

The risk (HRI) and its components (threat, vulnerability and response capacity) is determined; afterwards, these are translated into variables and indicators that are fed with available (or hypothetical) information in order to establish the baseline assessment of the system.

6.4.4 Implementation of the Work Plan

The activities are carried out through a program of training and dissemination of technology to and from the producers, under the FFSAE model.

6.4.5 Risk and Resilience Monitoring

Once adaptation and mitigation actions are started, the system is monitored to assess risk and resilience. A risk reduction (HRI) and an increase in resilience – with respect to the baseline data – will mean an improvement of the socio-environmental system against pests; if the system under management does not improve, the work plan should be carefully reviewed to make the necessary modifications to achieve the expected goals.

6.5 Holistic Risk Index: Case Study

In Chap. 5 it was shown that HRI is the cornerstone of HPM, since this index is the basis for decision-making. Given its importance, we will proceed to describe the methodology to estimate it. As an example, the data of 12 coffee producers from the Soconusco region of Chiapas, Mexico (who participated in a HPM study) will be used to determine the importance of organic coffee certification as a measure of adaptation to the uncertainty caused by low coffee prices, extreme weather and pest attack (Castellanos et al. 2015; Barrera et al. 2018; Barrera 2020).

At the field level the producers were distributed in pairs: one organic coffee certified and the other – its neighbor – not certified. In total, six pairs of producers (six certified and six non-certified) participated in the study. The first four couples (producers 1–8) were smallholder producers and the other two (9–12) were large producers.

The example presented here corresponds to the coffee leaf rust (*Hemileia vastatrix*) HRI estimation of these producers. The data was recorded in 2013, after the strong attack of the rust that occurred in Central America (Avelino et al. 2015) and southern Mexico (Barrera et al. 2013) during 2012. Coffee rust is a pathogenic fungus that attacks leaves of coffee trees; it infects with greater intensity some *Coffea arabica* varieties. The diseased leaves have orange pustules on their lower surface. After a strong attack, the leaves may fall. Rust has been known for a century and a half, and today it is found in virtually every coffee producing country in the world (Avelino and Rivas 2013).

6.5.1 Estimate

Socio-economic information was recorded through producer surveys, interviews and agroecological information through RAS in their plantations. The HRI was estimated following these six steps (Barrera et al. 2018):

(i) Definition of descriptors and indicators, and data collection
(ii) Determination of indicators importance
(iii) Relationship between variables and threat
(iv) Standardization of variables
(v) Calculation of risk components
(vi) HRI calculation

6.5.1.1 Definition of Descriptors and Indicators and Data Collection

Descriptors and indicators are defined for each HRI component as shown in Fig. 6.4. In this example, the threat (rust) was represented by the infection of coffee trees in the target plantation (refers to plantation where all the indicators were sampled) and the neighbor's plantation (in this plantation only rust infestation is considered); taking into account that it worked with pairs of producers (one certified and the other not), each plantation of one pair was target plantation and neighboring plantation of the other.

Vulnerability was made up of three variables: altitude of the coffee plantation (meters above sea level), rate of rust increase (relation between the infestation at the initial and final sampling time), and the area planted with coffee (hectares).

The response capacity was also composed of three variables: net profit from coffee sales, producer participation in an organic coffee certification program, and whether use fungicides to rust control.

Once the variables are defined, the indicators are sampled to obtain the data. The data can be obtained directly from plantations through the RAS or by using information from farmer or farmer organization. In the event that no real data is available for any indicator, estimated data can be used.

In this example, collected data from indicators in Fig. 6.4 are presented in Table 6.1. This table shows the original data of the 12 Chiapas producers who participated in the study, six certificates and six non-certified organic coffee producers. The data is ordered according to the components of the HRI: threat, vulnerability and response capacity.

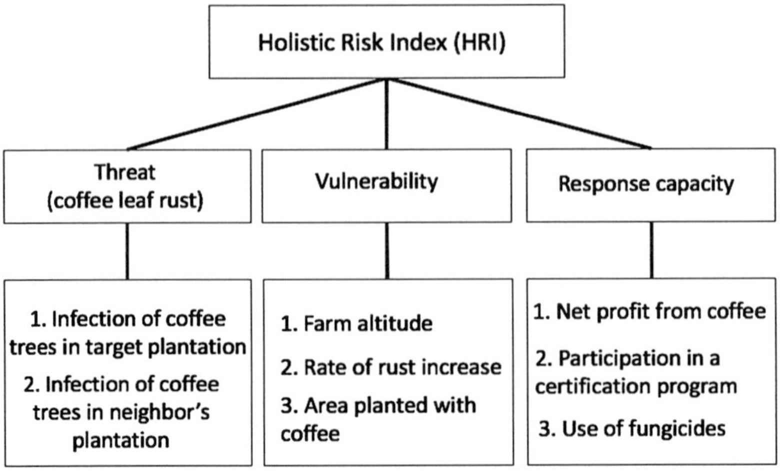

Fig. 6.4 Descriptors and indicators used to estimate the HRI of coffee producers to coffee leaf rust (*Hemileia vastatrix*). (from Barrera 2020) © 2020 From Holistic pest management by Juan F. Barrera. Reproduced by permission of Taylor and Francis Group, LLC

6.5.1.2 Determination of Indicator Importance

Considering that some indicators may be more important than others, it is desirable to determine the degree of their individual contribution in explaining the observed variability of the system under study (in this example the 12 producers). For this, it is recommended to use some multivariate method such as a factor analysis. In case of not having the means to do this statistical analysis, the weights or indicator importance can be assumed (as a last option it can be assumed that the indicators have the same weight).

If we apply a factor analysis to the data in Table 6.1, we get the result shown in Table 6.2. As can be seen, the indicators under study were grouped into two factors that in total explained 71% and 29% of the variance observed. The most important indicators grouped by factor 1 were: coffee planted area, net profit from coffee sale, and the use of fungicides; while the most important indicators grouped by factor 2 were: plantation infestation, altitude, rust increase rate and certification program (Table 6.2).

6.5.1.3 Relationship Between Indicators and Threat

Establishing the relationship between the indicators under study and the threat – in this case the coffee rust – is a fundamental aspect of HRI analysis. To establish these relationships, it is necessary to know the behavior of the threat under various situations. Preferably, the relationships established must be based on information from scientific research or empirical knowledge.

For example, it is known that personal factors such as age, schooling, family size and sex of farmers, and operational factors such as the size of the coffee land,

Table 6.1 Original data from 12 certified (6) and non-certified (6) organic coffee producers in the Soconusco region, Chiapas, Mexico. Year 2013

	Threat, leaves with rust (%)[b]		Vulnerability			Response capacity		
Producer number and producer pairs[a]	Producer's farm	Farm of the neighboring producer[c]	Altitude (masl)[d]	Rust increase rate[e]	Farm size (ha)[f]	Net income from coffee sales ($ MXN)[g]	Certification (organic coffee)[h]	Use of fungicides[i]
1. Certif01	42.7	33.1	1156	2.33	2.5	265.38	1	0
2. No Certf01	33.1	42.7	1128	1.32	2.5	233.23	0	0
3. Certif02	73.7	50.5	1109	2.57	3.5	2174.36	1	0
4. No Certif02	50.5	73.7	1099	5.48	2.5	180.00	0	0
5. Certif03	56.3	42.9	988	1.84	3.0	1642.31	1	0
6. No Certif03	42.9	56.3	982	1.47	2.0	132.69	0	0
7. Certif04	3.9	28.4	1329	0.58	3.0	2794.87	1	0
8. No Certif04	28.4	3.9	1366	0.91	3.5	−8.79	0	0
9. Certif05	20.7	45.5	710	31.07	300.0	200,001.00	1	1
10. No Certif05	45.5	20.7	711	1.39	42.5	200,001.00	0	1
11. Certif06	1.5	3.6	857	4.73	301.0	200,001.00	1	1
12. No Certif06	3.6	1.5	1191	0.72	287.0	200,001.00	0	1

Modified from Barrera et al. (2018)
[a]Certif = certified, No certif = not certified. [b]The coffee leaf rust infection can be expressed in several ways; here the percentage of leaves with rust per branch on each coffee plant was preferred. [c]Since the producers were paired (one being certified and the other not), each one was the other's neighbor. [d]masl = meters above sea level. [e]The rate of leaf rust increase was calculated from two samplings, one done in September and the other in December, 2013. [f]The producer pairs from 1 to 4 are smallholder producers; pairs 5 and 6 are large producers. [g]In the case of the large producers, the total earnings for the total produced coffee are hypothetical. [h]Certified producer = 1, non-certified producer = 0. [i]The producer uses fungicide against the leaf rust = 1, the producer does not use fungicide = 0

access to own machinery, receive technical assistance and participating in a farmer organization influence the acceptance of technology (Regalado López et al. 1996). If this information is not available, hypothetical relationships can be established.

Figure 6.5 is an example of the relationships of the indicators under study with rust. It can be seen that the relationships were linear and positive –more rust infestation as the indicator increased– or negative –less infestation. Only in one case, a curvilinear relationship of plantation altitude with rust infestation was established.

Table 6.2 Factor analysis of the original data from 12 certified (6) and non-certified (6) producers as organic coffee producers in the Soconusco region, Chiapas, Mexico. Year 2013

Indicators	Indicator key	Factor 1	Factor 2
% of rust infection on farm	Producer	−0.67	0.47
% of rust infection on the neighbor's farm	Neighbor	−0.61	0.64
Altitude (masl)	Alt	−0.46	−0.78
Rust increase rate	Rate	0.38	0.50
Coffee planted area (ha)	Size	0.90	0.03
Net income from coffee sales	Income	0.97	0.11
Certification program	Cert	0.07	0.19
Use of fungicides	Fungi	0.97	0.11
Proportion explained		0.71	0.29
Cumulative proportion		0.71	1.00

Modified from Barrera et al. (2018)

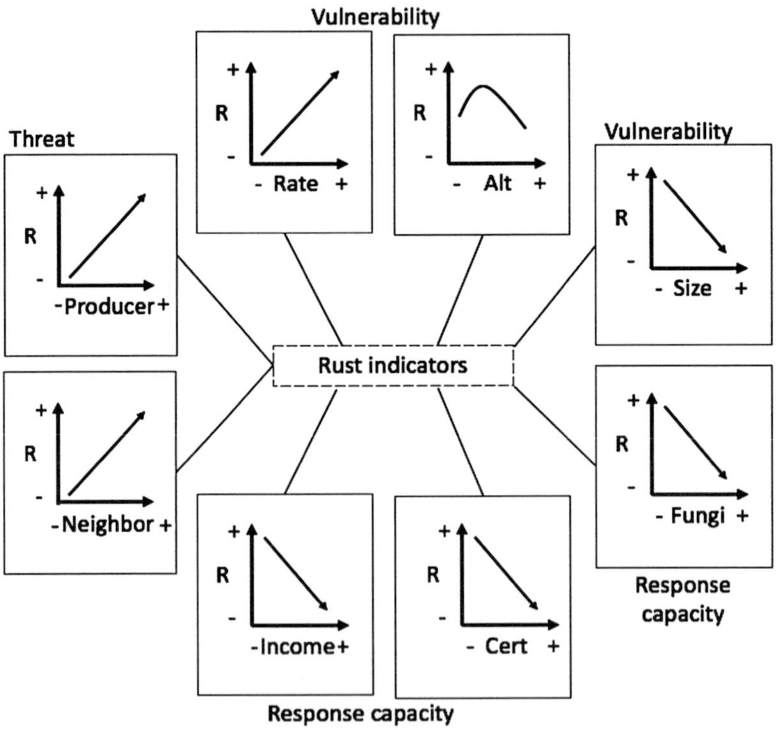

Fig. 6.5 Relationships between the indicators that integrate the components of HRI and the coffee leaf rust infestation. R = rust incidence, Producer = farm infestation, Neighbor = farm infestation, Rate = Rust increase rate, Alt = altitude, Size = area planted with coffee, Income = Net income from coffee sales, Cert = certified as organic, Fungi = use of fungicides. (Modified from Barrera et al. 2018)

6.5.1.4 Standardization of Indicators

Based on the relationships established between indicators and rust (Fig. 6.5), the indicators are standardized (converting the original data on a scale of 1 to 100). In the case of the threat, as can be seen in Fig. 6.6, as the rust infestation increased from <3.0% (very low infestation, 20 on the scale) to more than 60% (very high infestation, 100 on the scale), the value on the scale increased positively and linearly. A similar relationship was established for the rate of rust increase. The relationship was negative and linear for the other indicators. A different case was altitude, where vulnerability to rust had a value of 100 in the middle zone and lower values in the low (50), high (75) or very high (25) areas. In the coffee certification and use of fungicides, whose variables had only two options, there was a value of "0" for the "no" and "100" for the "yes". The results of the standardization are shown in Table 6.3.

6.5.1.5 Calculation of Risk Components

Standardized values (Table 6.3) are multiplied by the constants from factor analysis (Table 6.2) to include the weight of the indicators in the calculation of HRI components (Table 6.4). For example, in the case of the "Certif01" producer, the calculation is performed as follows:

$$\text{Threat}\,(\text{T}) = (\text{Producer} * \text{Constant Factor 2}) + (\text{Neighbor} * \text{Constant Factor 2})$$

$$\text{T} = (80 * 0.47) + (80 * 0.64) = 88.80$$

$$\text{Vulnerability}\,(\text{V}) = (\text{Alt} * \text{Constant Factor 2}) + (\text{Rate} * \text{Constant Factor 2}) + \text{Size} * \text{Constant Factor 1})$$

$$\text{V} = (75 * 0.78) + (100 * 0.5) + (75 * 0.9) = 176.00$$

$$\text{Response capacity}\,(\text{C}) = (\text{Income} * \text{Constant Factor 1}) + (\text{Cert} * \text{Constant Factor 2}) + \text{Fungi} * \text{Constant Factor 1})$$

$$\text{C} = (25 * 0.97) + (100 * 0.19) + (0 * 0.97) = 43.25$$

6.5.1.6 Risk Calculation

The data in Table 6.4 are transformed into percentages to calculate the HRI (Table 6.5). For example, in the case of the producer "Certif01", the calculation consists of multiplying the adjusted value of the component "x" (Table 6.4), times the factor analysis constant (Table 6.2), times 100, times the sum of the adjusted values of all the components (Table 6.4):

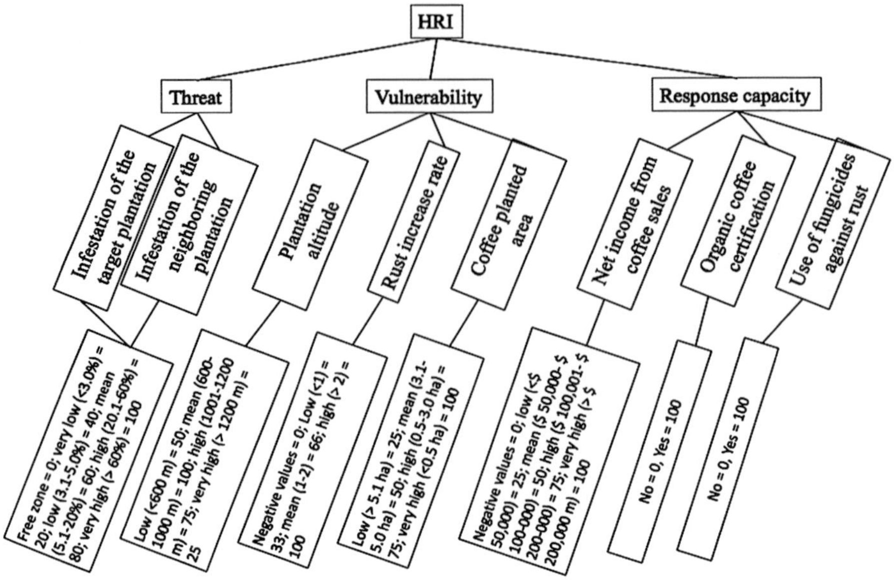

Fig. 6.6 Standardization of indicators related to the threat, vulnerability and response capacity to rust by coffee producers in the Soconusco region, Chiapas, Mexico. (Modified from Barrera et al. 2018)

$$\text{Threat} = 88.8 * 100 / 308.05 = 28.83\%$$
$$\text{Vulnerability} = 176.00 * 100 / 308.05 = 57.13\%$$
$$\text{Response capacity} = 43.25 * 100 / 308.05 = 14.04\%$$

Then, the HRI equation is applied, whose value in the case of the producer "Certif01" is calculated as follows (Table 6.5):

$$\text{HRI} = \left[\text{Threat}(\text{T}) + \text{Vulnerability}(\text{V})\right] / \text{Response capacity}(\text{C}).$$
$$\text{R} = (28.83 + 57.13) / 14.04 = 6.1$$

6.5.2 *Graphical Analysis*

Once the HRI and its components have been calculated, its graphic analysis is carried out using the THR and radial charts.

Table 6.3 Standardized data for the 12 certified (6) and non-certified (6) organic coffee producers in the Soconusco region, Chiapas, Mexico. Year 2013

	Threat, leaves with rust (%)[b]		Vulnerability			Response capacity		
Producer number and producer pairs[a]	Producer's farm	Farm of the neighboring producer[c]	Altitude (masl)[d]	Rust increase rate[e]	Farm size (ha)[f]	Net income from coffee sales ($ MXN)[g]	Certification (organic coffee)[h]	Use of fungicides[i]
1. Certif01	80	80	75	100	75	25	100	0
2. No Certf01	80	80	75	66	75	25	0	0
3. Certif02	50	80	75	100	50	25	100	0
4. No Certif02	80	50	75	100	75	25	0	0
5. Certif03	80	80	100	66	75	25	100	0
6. No Certif03	80	80	100	66	75	25	0	0
7. Certif04	40	80	25	33	75	25	100	0
8. No Certif04	80	40	25	33	50	1	0	0
9. Certif05	80	80	100	100	25	100	100	100
10. No Certif05	80	80	100	66	25	100	0	100
11. Certif06	10	20	100	100	25	100	100	100
12. No Certif06	20	10	75	33	25	100	0	100

Modified from Barrera et al. (2018)

[a]Certif = certified, No certif = not certified. [b]The coffee leaf rust infection can be expressed in several ways; here the percentage of leaves with rust per branch on each coffee plant was preferred. [c]Since the producers were paired (one being certified and the other not), one was the other's neighbor. [d]masl = meters above sea level. [e]The rate of leaf rust increase was calculated from two samplings, one done in September and the other on December, 2013. [f]The producer pairs from 1 to 4 are smallholder producers; pairs 5 and 6 are large producers. [g]In the case of the large producers, the total earnings for the total produced coffee are hypothetical. [h]Certified producer = 1, non-certified producer = 0. [i]The producer uses fungicide against the leaf rust = 1, the producer does not use fungicide = 0

Table 6.4 Adjustment of rust HRI component values of coffee producers in the Soconusco region, Chiapas, Mexico, using the indicator constants obtained through factor analysis. Year 2013

Producer	Threat	Vulnerability	Response capacity	Total
1. Certif01	88.80	176.00	43.25	308.05
2. No Certif01	88.80	159.00	24.25	272.05
3. Certif02	74.70	153.50	43.25	271.45
4. No Certif02	69.60	176.00	24.25	269.85
5. Certif03	88.80	178.50	43.25	310.55
6. No Certif03	88.80	178.50	24.25	291.55
7. Certif04	70.00	103.50	43.25	216.75
8. No Certif04	63.20	81.00	0.97	145.17
9. Certif05	88.80	150.50	213.00	452.30
10. No Certif05	88.80	133.50	194.00	416.30
11. Certif06	17.50	150.50	213.00	381.00
12. No Certif06	15.80	97.50	194.00	307.30

Modified from Barrera et al. (2018)

6.5.2.1 The Holistic Risk Triangle

The graphical representation of rust risk components for the 12 coffee producers in Table 6.5 is shown in Figs. 6.7 and 6.8. As can be seen, large producers had lower levels of threat and vulnerability and higher levels of response capacity than smallholders, which explains why the latter were located in the highest risk areas of THR (Fig. 6.7).

Although coffee certification was not decisive in the levels of risk to rust, it can be generally seen that the response capacity of non-certified smallholder producers was slightly lower (0.7–9.0%, Table 6.5) than that of certified smallholder producers (8.3–20%, Table 6.5). In terms of risk, non-certified smallholder producers had risks greater than 10.0, while the risk of certified smallholder producers was equal to or less than 6.2 (Table 6.5).

In the case of large producers, certification was even less important to reduce the risk to rust, however, it can be seen that producers who were certified had slightly lower threat levels, although they were more vulnerable and had less response capacity than non-certified large producers (Fig. 6.7).

From the 12 evaluated producers, it can be inferred that the vast majority of Soconusco region producers, presented medium and high risks to rust, particularly smallholder producer (Fig. 6.8a). Statistically speaking, smallholder producers made a significantly compact and different group than that of large producers (Fig. 6.8b).

Table 6.5 Calculation of rust HRI[a] of coffee producers in the Soconusco region, Chiapas, Mexico, based on the percentage value of its components. Year 2013

Producer	Threat (%)	Vulnerability (%)	Response capacity (%)	Risk
1. Certif01	28.8	57.1	14.0	6.1
2. No Certif01	32.6	58.4	8.9	10.2
3. Certif02	27.5	56.5	15.9	5.3
4. No Certif02	25.8	65.2	9.0	10.1
5. Certif03	28.6	57.5	13.9	6.2
6. No Certif03	30.5	61.2	8.3	11.0
7. Certif04	32.3	47.8	20.0	4.0
8. No Certif04	43.5	55.8	0.7	148.7
9. Certif05	19.6	33.3	47.1	1.1
10. No Certif05	21.3	32.1	46.6	1.1
11. Certif06	4.6	39.5	55.9	0.8
12. No Certif06	5.1	31.7	63.1	0.6

Modified by Barrera et al. (2018)
[a]HRI = (threat + vulnerability)/response capacity

6.5.2.2 Radial Charts

The HRI and its components can be analyzed in more detail by representing the indicators that integrate each risk component in radial charts such as those shown in Fig. 6.9. This figure compares certified producers against non-certified producers, both grouped (top chart) and in pairs. In the first case, with exception of the indicator that refers to coffee certification, the indicators of both groups are quite similar. However, the graphs that compare couples of producers show important differences in the indicator value in some cases (e.g., couples 4 and 6).

In summary, the analysis of rust risk through the HRI allows us to infer that the smallholder coffee producers considered in this study –in general the smallholder producers of Soconusco– are at very high risk of suffering damage from this disease. In addition, it shows that its resilience towards rust is low. The situation is different for large producers, who had lower risks towards rust. Farmers certified as organic coffee producers do not get rid of this threat, although in general terms the analysis indicated that they have a better response capacity.

In order to reduce threat and vulnerability and increase the capacity for response and resilience towards rust, the work plan must include actions that, on the one hand, reduce the incidence of rust, and on the other, improve producers' incomes. For this, it is essential to renew coffee plantations, diversify production and markets, transform the product to add value, and consolidate the producer organization, among other actions.

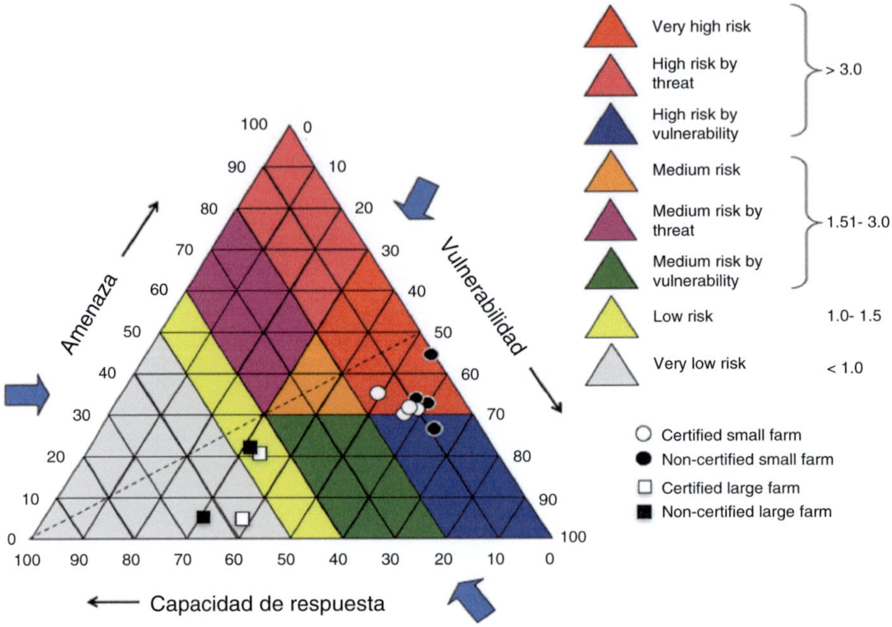

Fig. 6.7 Holistic Risk Triangle to rust HRI components for 12 certified and non-certified organic coffee producers in the Soconusco region, Chiapas, Mexico. Year 2013. (from Barrera 2020) © 2020 From Holistic pest management by Juan F. Barrera. Reproduced by permission of Taylor and Francis Group, LLC

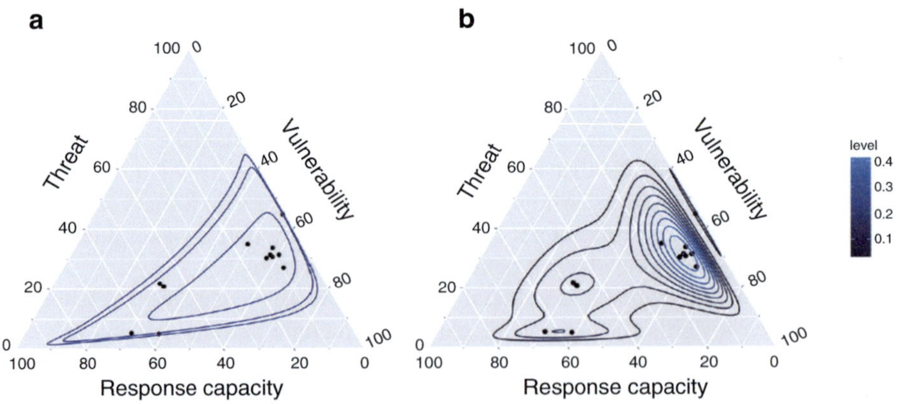

Fig. 6.8 Confidence intervals (**a**) and probability density function (**b**) of the Holistic Risk Triangle to rust for 12 certified and non-certified organic coffee producers in the Soconusco region, Chiapas, Mexico. Year 2013. (from Barrera 2020) © 2020 From Holistic pest management by Juan F. Barrera. Reproduced by permission of Taylor and Francis Group, LLC

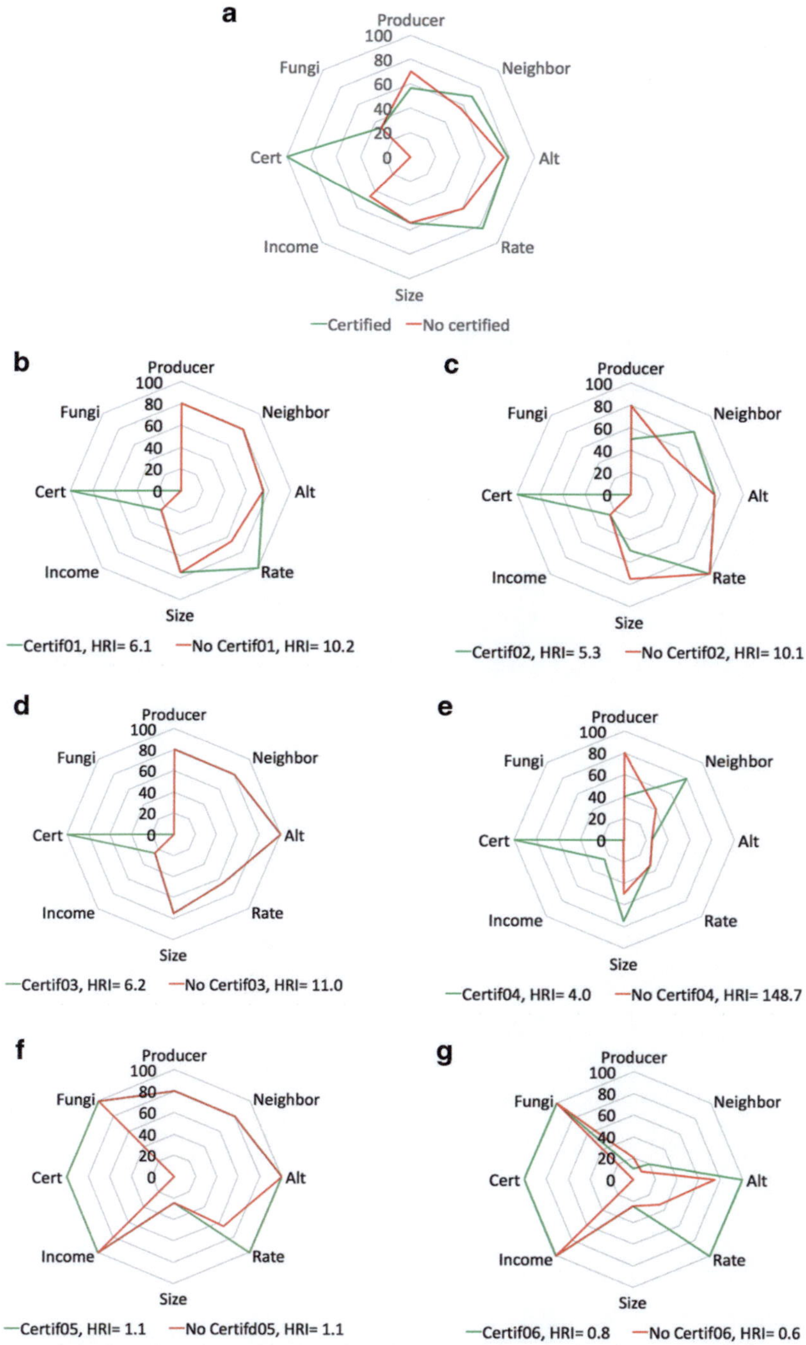

Fig. 6.9 Radial charts that compare the rust risk components for certified producers against non-certified producers, both grouped (**a**) and in pairs (**b–g**) in the Soconusco region, Chiapas, Mexico. Year 2013. Producer = farm infestation, Neighbor = farm infestation, Rate = Rust increase rate, Alt = altitude, Size = area planted with coffee, Income = Net income from coffee sales, Cert = certified as organic, Fungi = use of fungicides. (Modified from Barrera et al. 2018)

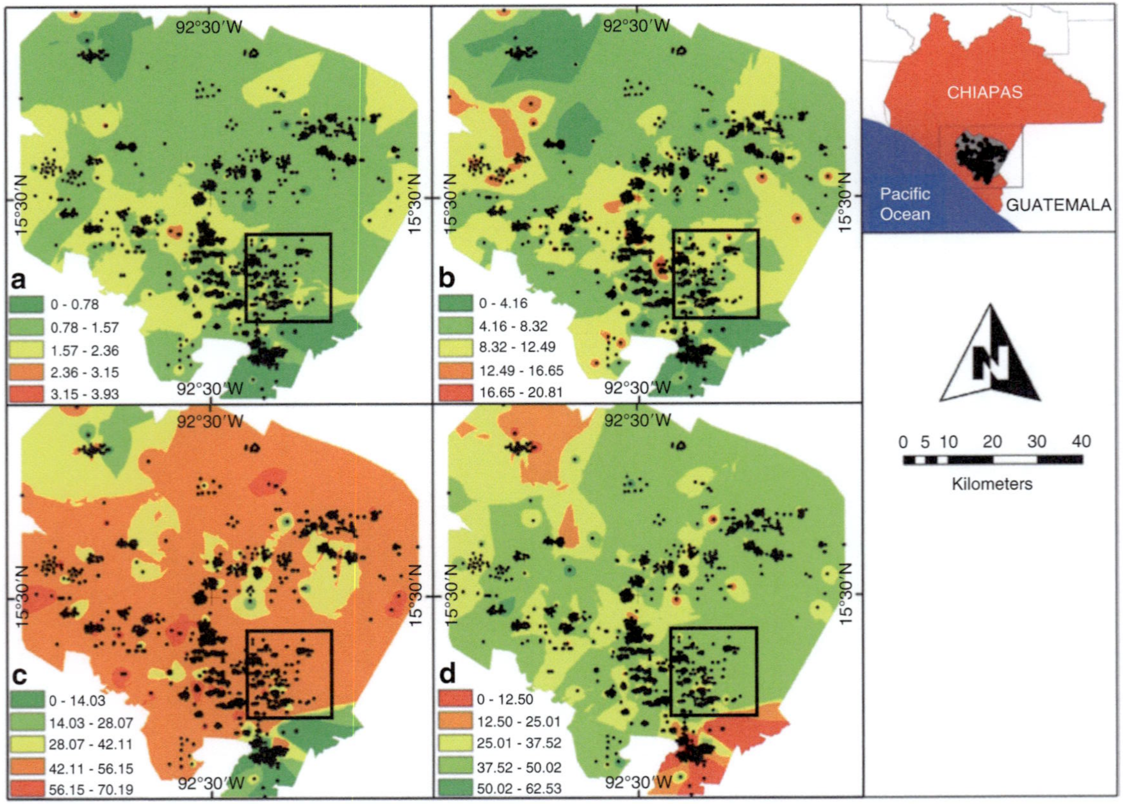

Fig. 6.10 Geographical representation of the Holistic Risk Index (**a**) and its components: the threat or rust (**b**); the vulnerability or altitude of the farm (**c**); and the response capacity or net profit (**d**) of around 2500 producers associated with an organic coffee organization in Chiapas. Years 2014–2015. The green colors in **a**, **b** and **c** indicate low values, and in **d** indicate high values. (from Barrera 2020) © 2020 From Holistic pest management by Juan F. Barrera. Reproduced by permission of Taylor and Francis Group, LLC

6.5.2.3 Maps

If a georeferenced database with enough producers is available to perform geographic analysis, maps representing the HRI and its components can be developed. To illustrate this case, four maps that integrate Fig. 6.10 are presented, prepared from a database of around 2500 producers associated with an organic coffee organization in Chiapas. The maps represent the HRI (Fig. 6.10a) and its components: the threat or rust (Fig. 6.10b); the vulnerability or altitude of the farm (Fig. 6.10c); and the response capacity or net gain (Fig. 6.10d).

When observing Fig. 6.10a, b, which represent the HRI and the threat respectively, it can be seen that the colors of the maps – intensity of the variables – vary with respect to each other; the difference is given because the estimation of the HRI considers information about the threat (Fig. 6.10b) the vulnerability (Fig. 6.10c) and the response capacity (Fig. 6.10d). In other words, with the HRI, better decisions are made for rust control, as decisions are made with more information. If a more specific region of these maps is observed, for example, the one indicated in the small boxes, it will be noted that the threat shows greater intensity (greater area covered by yellow color) than the HRI; that is, while the threat – and vulnerability – indicates the need to act against rust, the HRI points out that it is not necessary to hurry so much because most of the producers located in the small box on the map have a good response capacity.

6.6 Conclusion

In this chapter, a guide to implement the HPM was presented, from assessing the socio-environmental system to determine its problems, to assessing the effectiveness of adaptation and mitigation actions to reduce its risk and increase its resilience to pests. It is not a prescriptive or exhaustive guide, but illustrative and simple to facilitate the management of pests under a holistic approach. Integrating a working group with the participation of farmers, academics and facilitators is one of the most important actions, since this group will be responsible for implementing the HPM. The cornerstone of the HPM, because of its importance in decision-making, is the HRI. This chapter illustrates the steps to be taken to estimate it and presents a case with coffee rust to study its application.

References

Altieri, M. A. (1993). Ethnosience and biodiversity: Key elements in the design and of sustainable pest management systems for small farmers in developing countries. *Agriculture, Ecosystems and Environment, 46*(1–4), 257–272.

Altieri, M. A. (1994). *Biodiversity and pest management in agroecosystems*. New York: Hayworth Press.
Altieri, M. A., & Nicholls, C. I. (2003). Ecologically based pest management: A key pathway to achieving agroecosystem health. In D. J. Rappoport, W. L. Lasley, D. E. Rolston, et al. (Eds.), *Managing for healthy ecosystems* (pp. 999–1010). Boca Raton: Lewis Publishers.
Altieri, M. A., & Nicholls, C. I. (2007). *Biodiversidad y manejo de plagas en agroecosistemas*. Barcelona: Icaria Editorial.
Altieri, M. A., & Trujillo, J. (1987). The agroecology of corn production in Tlaxcala, Mexico. *Human Ecology, 15*(2), 189–220.
Andow, D. A. (1991). Vegetational diversity and arthropod population response. *Annual Review of Entomology, 35*, 561–568.
Avelino, J., & Rivas, G. (2013). *La roya anaranjada del cafeto*. https://hal.archives-ouvertes.fr/hal-01071036. Accessed 28 Oct 2019.
Avelino, J., Cristancho, M., Georgiou, S., et al. (2015). The coffee rust crises in Colombia and Central America (2008–2013): Impacts, plausible causes and proposed solutions. *Food Security, 7*, 303–321.
Barrera, J. F. (2020). Holistic pest management. In D. Perez-Staples, F. Diaz-Fleischer, P. Montoya, & M. T. Vera (Eds.), *Area-wide management of fruit fly pests* (pp. 275–293). Boca Raton: CRC Press, Taylor & Francis Group.
Barrera, J. F., Avelino, J., Huerta, G., et al. (2013). La roya del café: Crónica de una devastación anunciada. *Ecofronteras, 49*, 22–25.
Barrera, J. F., Valle, J., Gómez, J., et al. (2018). *Manejo holístico de plagas en zonas cafetaleras: Concepto y método*. Universidad Autónoma Chapingo- Centro Nacional de Investigación, Innovación y Desarrollo Tecnológico del Café-El Colegio de la Frontera Sur. Mexico, 42 p.
Castellanos, E., Díaz, R., & Barrera, J. F., et al. (2015). *Small coffee producers and adaptive options for a changing climate: The risks and challenges of certification for ecosystem services*. Coffee and Global Changes Project (Project SGP-CRA 2060). Final Technical Report. Guatemala, América Central, 53 p.
Chapin, F. S., III, Zavaleta, E. S., Eviner, V. T., et al. (2000). Consequences of changing biodiversity. *Nature, 405*, 234–242.
de Camino, R., & Müller, S. (1993). *Sostenibilidad de la agricultura y los recursos naturales: bases para establecer indicadores*. Instituto Interamericano de Cooperación para la Agricultura/ Proyecto IICA/GTZ (Serie Documentos de Programas No. 38). San José, Costa Rica, 134 p.
Mojica, S. F. (1991). *La prospectiva; técnicas para visualizar el futuro*. Bogotá: Fondo Editorial Legis.
Mojica, F. J. (2004). *El modelo prospectivo llevado a la práctica. Primera versión. Convenio Andrés Bello (CAB)*. Serie: Documentos de Ciencia, Tecnología e Innovación de los países del CAB. Bogotá, Colombia, 178 p.
Perrings, C. (2005). Mitigation and adaptation strategies for the control of biological invasions. *Ecological Economics, 52*, 315–325.
Regalado López, J., Jiménez Sánchez, L., Laird, R. J., et al. (1996). Factores asociados a la utilización de la tecnología de alta productividad entre productores de maíz de temporal. *Agrociencia, 30*(1), 139–145.
Wyss, E., Luka, H., Pfiffner, L., et al. (2005). *Approaches to pest management in organic agriculture: A case study in European apple orchards*. Organic-Research.com, May 2005: 33N–36N. https://www.researchgate.net/publication/262002057_Approaches_to_management_in_organic_agriculture_a_case_study_in_European_apple_orchards. Accessed 24 Oct 2019.
Zehnder, G., et al. (2007). Arthropod pest management in organic crops. *Annual Review of Entomology, 52*, 57–80.

Chapter 7
Holistic Science in Pest Management: Winds of Change

It is time for IPM theorists and practitioners to declare mission accomplished with the IPM… Today the challenge is to seek a new orientation, with a new way of planning the path of plant protection; that includes its legacy, which has been a paradigm of the last 50 years, but that transcends it.
Keith L. Andrews. X International Congress of Integrated Pest Management and Agroecology (Translated from: Andrews KL (2006) Prólogo. In: Castillo A, Gómez J, Morales H, Toledo J, Jarquín R (eds) Resúmenes del X Congreso Internacional de Manejo Integrado de Plagas y Agroecología. 27 al 29 de septiembre de 2006. Tapachula, Chiapas, México).

Abstract In this final chapter, the characteristics of Integrated Pest Management (IPM) – the current pest management paradigm– and Holistic Pest Management (HPM) – the alternative paradigm – are confronted in a *vis-à-vis* comparison that emphasizes the differences between both paradigms. Subsequently, cases are presented where the Holistic Risk Index (HRI) – the decision-making cornerstone of HPM – has been used to study threats in other contexts than pests. The most significant challenges of the HPM are mentioned in the last part of the chapter: challenges to transform a society with pesticides into a fairer, more environmentally friendly one that promotes and supports economically viable socio-environmental systems.

Keywords Integrated Pest Management · Holistic Pest Management · Paradigm · Holistic Risk Index · Decision-making

J. F. Barrera, *Beyond IPM: Introduction to the Theory of Holistic Pest Management*, Sustainability in Plant and Crop Protection 15,
https://doi.org/10.1007/978-3-030-43370-3_7

7.1 Introduction

Through the preceding chapters, the arguments that justify proposing a new paradigm to deal with pests in socio-environmental systems have been presented. The new paradigm, named Holistic Pest Management (HPM), is characterized by its holistic approach and being centered on people. These characteristics make the HPM a desirable alternative paradigm to Integrated Pest Management (IPM), the current paradigm.

In a recent publication, Stenberg (2017) suggested that IPM is still far from being investigated or applied efficiently because – according to him – a holistic science of IPM has probably not yet emerged. We categorically affirm that the lack of a holistic science has been the main cause of IPM paradigm anomalies. According to Dr. Keith Andrews's metaphor,[1] it is no longer possible to think of IPM as a building with a facade and a floor, much less as an isolated building (Fig. 7.1).

Therefore, it is clear to us that without the holistic approach, the implementation of IPM will be a strategy of only good intentions in most cases. Indeed, it is illusory to think that effective pest management is achieved through isolated and unrelated actions from the rest of the system, especially in cases with low-income farmers, in large areas and in the long term.

Unlike the IPM in concept, HPM also differs in the methods, as we will detail later. HPM is a strategy to deal with complexity, that is, with the networks of interactions that exist in real systems that Capra (1998) called the "web of life". In addition, the systems approach followed by HPM makes it possible to provide the appropriate adjustments to study and manage the risk posed by threats other than pests, which speaks of its flexibility and versatility. It is not a question of renaming or redefining IPM, as Royer et al. (1999) say when they refer to some alternatives to IPM, but indeed to replace it with HPM.

This final chapter has three objectives: first, to point out the most important differences between IPM and HPM; second, to give examples where the Holistic Risk Index (HRI) –the cornerstone in HPM decision-making system – has been used to investigate other threats in other contexts; and third, to glimpse the challenges of HPM in the face of coming changes.

7.2 Comparison Between IPM and HPM

The preceding chapters highlighted strong differences between IPM and HPM; however, this section presents a *vis-à-vis* comparison between them to clarify the differences between the two approaches (Table 7.1), that maybe established considering the following seven aspects (Barrera 2020):

[1] Keynote speaker, X Congreso Internacional de Manejo Integrado de Plagas y Agroecología. September 27–29, 2006. Tapachula, Chiapas, Mexico.

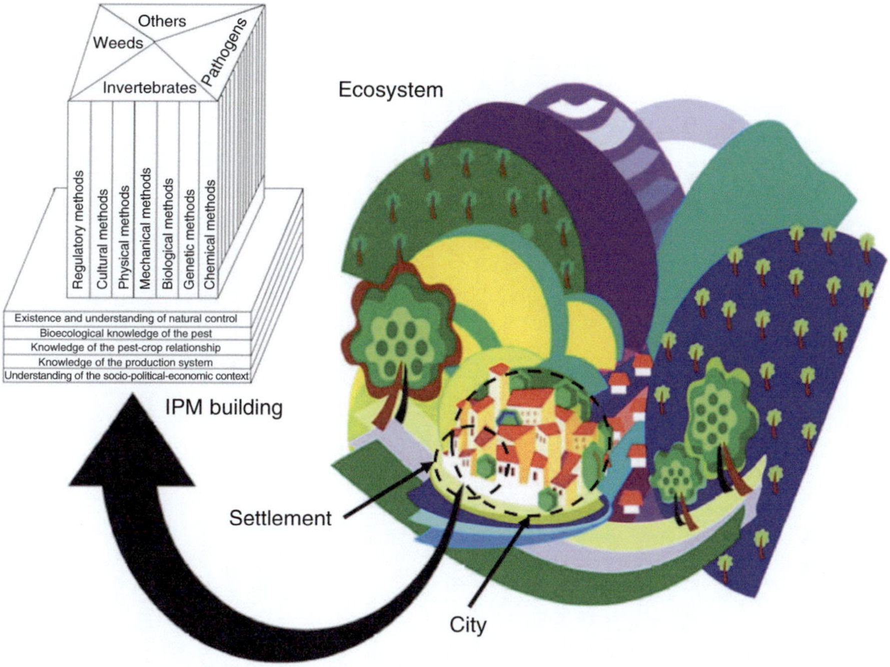

Fig. 7.1 IPM has been metaphorically represented as a building whose foundations are the existence and understanding of natural control, knowledge of the pest, the pest-crop relationship and the production system, and understanding of the socio-political-economic context. In this imaginary building (upper left figure, Andrews and Quezada 1989), the walls represent the control methods (regulatory, cultural, physical, mechanical, biological, genetic and chemical) and the roof tiles are the groups of pest organisms (invertebrates, pathogens, weeds, etc.). However, IPM should not be thought of as a building with a facade and a floor, much less as an isolated building. Under HPM, pest management is only one element (building) of a subsystem (settlement) that forms and is part of a complex system (city). (Modified from Barrera 2009)

(i) Priority of the paradigm
(ii) Participation of the farmer
(iii) Type of response promoted by the paradigm
(iv) The decision-making system
(v) Strategy of the paradigm
(vi) More favorable production systems for paradigm application
(vii) Scale where the paradigm operates best

A summary of each of these aspects is presented below.

Table 7.1 Comparison between IPM and HPM approaches in socio-environmental systems (from Barrera 2020).

Approach	IPM	HPM
Priority	Management focused on the pest. IPM emphasizes the pest management through harmonious integration of various control methods with the least economic, environmental and social impact[a]. Even, the "biointensive IPM", whose approach is more holistic, complex and sensitive to change, is based on understanding the ecology of the pest[b]. However, by favoring attention over pests, attention to the farmer usually becomes secondary; in general, this situation is taken to the extreme in area-wide management programs[c], particularly in cases of pest eradication[d], where people living in the region where these programs operate are rarely taken into account[e].	Management focused on improving the well-being of the farmer. The ultimate purpose should not be the pest, not even the crop, but the viability of rural life as a whole[f]. The root of HPM comes from "Holistic Management", which proposes to manage natural resources, promote biodiversity, improve production and generate financial strength to improve the quality of life while conserving the environment[g]. For this, it is necessary to resolve economic, social and environmental aspects that limit the lifeways[h] of farmers. As a consequence of improving the well-being of farmer (i.e. increased income), the conditions for carrying out pest management are improved[i].
Participation	The farmer almost always acts as a receiver. The farmer is a user of the information on pest management. In general, processes follow a top – down (i.e. government, technicians – farmers) pathway[j]. Typically, technicians employed by the government bring technological packages to farmers[k].	The farmer receives and contributes. Farmer participates in information generation process for pest management[l]. Emphasis is placed on action-research processes[m] and training through Farmer Field Schools[n]. Knowledge and perceptions of farmers play an important role in pest management[o].
Response	Reactive response. It is based on management of symptoms caused by pests (i.e. damage, population outbreaks). The focus on symptoms leads to this question: What should I do to control the pest?[p] Answering this question promotes the use and excessive use of pesticides (i.e. chemical, biological, genetically modified organisms)[q]. A reactive response leads to a reductionist approach.	Preventive response. It is based on managing the causes that trigger pests. Understanding the causes that give rise to a problem of pests leads to the question: Why is the pest a pest?[r] Answering this question requires knowing the structure and understanding the function and dynamics of the system, very much in the manner proposed by the ecologically based pest management[s]. A preventive response leads to a holistic approach.

Decision-making	Decision making is based on action thresholds (Economic Injury Level and Economic Threshold). The Economic Injury Level (EIL), defined as the lowest pest population density that will cause economic damage, is calculated with information on cost of control (C), value of the product in the market (V), losses attributable to the pest (I, D) and the cost of pest control (K), through this equation: $EIL = C/VIDK$. The Economic Threshold (ET) is the density at which control measures should be determined to prevent an increasing pest population from reaching the EIL. In other words, ET is the operational criterion that determines whether or not a management action against a pest is required. These thresholds, most useful in the application of therapeutic practices, are widely recognized as the most important concepts of IPM[t].	Decision making is based on risk. The risk, which can be defined as the likelihood of an adverse event and the magnitude of the consequences[u], is obtained through the Holistic Risk Index (HRI). The components of HRI are the threat (A), the vulnerability (V) and the response capacity (C), and is calculated with $HRI = (A + V)/C$[v]. The threat may be the occurrence of one (or more) pest(s) acting on certain conditions of vulnerability. Vulnerability is integrated with the characteristics of producers and farms, which determine their degree of exposure to a threat. The response capacity refers to the attributes and mechanisms of producers (and other stakeholders), to reduce the risks and survive, resist and recover from the damage. In this approach, the action thresholds are one of several indicators used to calculate the components of HRI.
Strategy	Search for "silver bullet" products and tactics. This begins with the idea that each problem has a unique and effective solution. These solutions are temporary. When the solution fails, the replacement is sought. This strategy leads to input substitution, that is, to permanently search for the silver bullet[w].	Increase resilience. This approach promotes the health of the system. To a large extent, the resilience of the system is increased by reducing the risk to the threat, while reducing vulnerability and increasing the response capacity of the farmer and other stakeholders involved in the system[x].
Production system	Modern industrial agriculture. IPM tends to be more efficient in monoculture production systems, generally for export or conventional markets[y], situation that occurs mostly in developed countries (IPM emerged in the context of US agriculture)[z]. These systems depend to a large extent on using large quantities of external inputs (pesticides and fertilizers) to produce large volumes of products per unit area. Therefore, IPM is considered as an essential tool to reduce dependence on pesticides[aa].	Agroecological agriculture. HPM tends to be more efficient in diversified systems, either for the export of products in special markets (i.e. organic, fair trade) or self-consumption. These systems with greater diversification of species –less prone to pest invasion and outbreaks[ab] – make use of few external inputs because they strengthen natural processes (i.e. water and nutrient cycles, beneficial fauna and flora) and prioritize the conservation of natural resources and quality over quantity[ac].

(continued)

Table 7.1 (continued)

Approach	IPM	HPM
Scale	From small to large scale. However, the farm is almost always the unit in which decisions are made[ad]. In some cases, it can be implemented in large areas with participation of many individual farms (area-wide approach[ae], i.e. Mediterranean fruit fly program[af]).	From small to large scale. The size of the scale does not matter in the implementation. However, to be implemented in large areas it is desirable to do it in coordination with farmer organizations[ag].

[a]Bottrell (1979), Kogan (1998), Norris et al. (2003), Prokopy y Kogan (2003), and Gray et al. (2009); [b]Benbrook (1996); [c]Lindquist (2000) and Hendrichs et al. (2007); [d]Walters et al. (2009); [e]Klassen (2000); [f]Levins (2007); [g]Savory y Butterfield (1999); [h]Herrera et al. (2019); [i]Barrera (2006); [j]Thrupp (1996); [k]Morse y Buhler (1997); [l]Thrupp (1996); [m]van de Fliert y Braun (2002); [n]Braun et al. (2000); [o]Bentley y Andrews (1996), Segura et al. (2004), Liebig et al. (2016), and Munyuli et al. (2017); [p]Lewis et al. (1997); [q]Hokkanen (2015); [r]Lewis et al. (1997); [s]NRC (1996) and Menalled et al. (2004); [t]Stern et al. (1959), Higley y Pedigo (1996), and Higley y Peterson (2009); [u]Griffin (2012); [v]Barrera et al. (2007) and Barrera et al. (2011); [w]Altieri y Nicholls (2005); [x]Barrera et al. (2007), Montalba et al. (2013), Henao-Salazar (2013), and Machado-Vargas et al. (2018); [y]Morse (2009); [z]Parsa et al. (2014); [aa]Benbrook (1996) and Aselage y Johnson (2009); [ab]Levins (2007); [ac]Nicholls et al. (2016); [ad]Levins (2007); [ae]Lindquist (2000) and Hendrichs et al. (2007); [af]Enkerlin et al. (2017); [ag]Barrera et al. (2017)

7.2.1 Priority of the Paradigm

The producer is not the priority in IPM. By personal experience it was shown that the fundamental issue is meeting the needs of producer to live well, as stated in HPM. Because IPM does not care about the producer, they do not care about IPM and, therefore, is not careful in its implementation.

7.2.2 Farmer Participation

In IPM the producer is not the one who makes the decisions. In IPM the producer neither defines its priorities nor generates the information it needs. On the contrary, the participation of producer in decision-making is sought in HPM and the activities are proposed to overcome the problems he defines.

7.2.3 Type of Response Promoted by the Paradigm

IPM is essentially reactive. Since IPM emerged in response to reducing the excessive and improper use of pesticides, its philosophy is reaction-oriented; this characteristic has led to little attention on prevention, a basic aspect of HPM.

7.2.4 The Decision-Making System

Decision-making in IPM is based on action thresholds. In IPM, decisions are made according to population density of the pests and the damage they cause. In overlooking other important components of the system, the IPM approach is reductionist when compared to the risk-based approach of the HPM and its components: the threat, vulnerability and response capacity.

7.2.5 Paradigm Strategy

IPM very often uses "silver bullets" pesticide products. Resorting to the silver bullets provides guaranteed solutions regardless of the circumstances. However, since silver bullets focus on the symptoms and not the causes, they rarely solve the problems caused by pests. Silver bullets trigger endless negative processes such as pesticide substitution. On the contrary, the HPM focuses on solving problems from their causes, posing as the principle of action to answer the question asked by Lewis et al. (1997) "*Why is the pest a pest*?"

7.2.6 More Favorable Production Systems for the Application of the Paradigm

IPM is oriented to situations of intensive agriculture and production for the market. Therefore, IPM methods do not integrate well with agroecological methods. Farmers focused on intensive production forget about production for self-consumption. As for the HPM, its actions are inspired by processes that go from the local to the global, a strategy that emphasizes food sovereignty.[2]

7.2.7 Scale Where the Paradigm Operates Best

IPM operates primarily on a small scale. Operating at the farm level makes it difficult to implement IPM at the community or regional level. HPM can be applied at the farm level, but one of its virtues is its application at regional level.

7.3 Application of HRI in Other Threats and Contexts

After the work of Barrera et al. (2007, 2011), HRI – the cornerstone of the HPM in decision-making – has generated interest in diverse contexts and with different threats in Latin America (Table 7.2).

Among the threats, in addition to pests in coffee plantations such as *Hypothenemus hampei* (Barrera et al. 2007, 2011; Castro-Triana 2018), *Hemileia vastatrix* (Reyes-Robles et al. 2017) and *Eurhizococcus colombianus* (Meneses-Ospina 2015), the HRI has been applied to investigate the impact of droughts (Montalba et al. 2013, 2015), floods (Carpio Sacoto and Carpio Sacoto 2014), over grazing (Gazzano Santos 2014; Gazzano et al. 2015, 2016; Gazzano and Achkar 2015, 2016), climate intensity (Henao-Salazar 2013; Álvarez Morales 2015), water availability and coffee prices (Machado-Vargas 2017, Machado-Vargas et al. 2018), on resilience and sustainability of various socio-environmental agricultural, livestock systems and conservation areas.

These studies showed the interest of having methodologies that help to understand complex systems, and facilitate decision-making based on risk analysis. These new ways of seeing the world – with participation of people – need other approaches and tools. As Scoones (2017) points out, complexity approaches have become an important tool for the study of rural systems. In this sense, HRI offers the possibility of moving from disciplinary to interdisciplinary and transdisciplinary approaches.

[2] Food sovereignty "*emphasizes ecologically adequate production, distribution and consumption, socio-economic justice and local food systems as a means to combat hunger and poverty, guaranteeing sustainable food security for all people*" (Anderson 2018).

Table 7.2 Cases where the Holistic Risk Index (HRI) has been used to investigate different threats in varying contexts

Context	Threat(s)	Country	References
Holistic pest management in coffee plantations, coffee berry borer	Infestación de *Hypothenemus hampei* (Ferrari) (Coleoptera: Curculionidae)	México	Barrera et al. (2007, 2011)
Resilience, water resources, rural and indigenous communities	Drought intensity	Chile	Montalba et al. (2013, 2015)
	Drought frequency		
	Loss of yields or productivity due to drought		
Resilience, socio-ecological systems, extreme weather events	Climate events intensity	Colombia	Henao-Salazar (2013)
	Duration		
	Frequency		
	Damage levels		
Social and ecological adaptation strategies, climate change, farmers	Floods	Equator	Carpio Sacoto y Carpio Sacoto (2014)
	Heavy rains		
	Landslides		
	Droughts		
Cattle producers, protected areas, estuaries	Decrease in grazing lands	Uruguay	Gazzano Santos (2014), Gazzano et al. (2015, 2016), and Gazzano y Achkar (2015, 2016)
	Negative socioeconomic and ecological impact of the basin		
	Insecurity for livestock farming		
Agroecological sustainability, agricultural production systems	Climate event	Mexico	Álvarez Morales (2015)
	Intensity		
	Frequency		
	Level of damage		
Agroecological management of the ground pearl in blackberry	Infestation of *Eurhizococcus colombianus* Jakubsky (Hemiptera: Margarodidae)	Colombia	Meneses-Ospina (2015)
Holistic pest management in coffee plantations, coffee rust	Infestation of *Hemileia vastatrix* Berk. & Br. (Basidiomycota: Uredinales)	Mexico	Barrera et al. (2017)
Holistic pest management in coffee plantations, coffee rust	Infestación de *Hemileia vastatrix* Berk. y Br. (Basidiomycota: Uredinales)	Honduras	Reyes-Robles et al. (2017)
Socioecological resilience of small coffee production	Water availability due to rainfall	Colombia	Machado-Vargas (2017), Machado-Vargas et al. (2018)
	Coffee price fluctuations		
Repellence and attraction in the coffee berry borer	Infestación de *Hypothenemus hampei* (Ferrari) (Coleoptera: Curculionidae)	Colombia	Castro-Triana (2018)

Modified from Barrera (2020).

7.4 Challenges in the Post IPM Era

In the opinion of Drucker (1993), every so often society changes its worldview; that is, society restructures itself to produce new values and other political and social order, with other arts and institutions.

In the field of pest management, have we been experiencing a transformation like the one pointed out by Drucker (1993) that has gone unnoticed by most? If that were the situation, what interests move us and what difficulties prevent us from seeing the change that is needed? In any case, are we prepared to change the way we think and act? In a scenario of change, what are the challenges to transform a mentally predisposed society to use pesticides – as Waage (1992) rightly pointed out – towards a more equal society, responsible and committed to its environmental and socioeconomic environment?

For HPM to operate and contribute to the development of this ideal society, farmers, academics and consumers are required to take on the following challenges:

(i) Environmental challenges:

Know the agroecological bases necessary to understand the processes and interactions between the elements of the agroecosystem, in order to design productive, biodiversity and multifunctional socio-environmental systems that operate with self-generated inputs, reduce pests outbreaks and other negative impacts on natural resources. In other words, "produce agricultural products while natural resources are conserved" becomes a fundamental challenge for HPM.

(ii) Economic challenges

Know the necessary administration bases for the farm effective management, emphasizing the quality and safety of processes and products, in particular avoiding the indiscriminate use of pesticides and other agrochemicals, with a view to product commercialization in differentiated markets that improve income and promote people's values. "The profitability of the business", without relying on agrochemicals, is a vitally important challenge in the implementation of HPM.

(iii) Social challenges

Know the livelihoods of farmers and their workers in order to help strengthen social organization and promote self-management, both to facilitate the social appropriation of scientific and technological knowledge and to promote public policies that improve the well-being of rural populations. Knowing how people make a living, and what their expectations and values are, is a very important challenge for the proper functioning of HPM.

Faced with these challenges, a change in attitude and focus is fundamental; the attitude should be in favor of people's well-being, and the focus should be directed to the socio-environmental system as a whole. In other words, it is necessary to think – and even more important – to act holistically; only in this way will progress be made towards what Morin (2000) called "the reform of thought".

7.5 Conclusion

Between IPM (current paradigm) and HPM (alternative paradigm) there are notable differences regarding their priorities, farmer participation, the responses they promote, the decision-making system, the strategies they follow, the production systems where they have the greatest application, and the geographical scale where they operate best. These differences highlight two characteristics of HPM: it is sociocentric and holistic. These characteristics mean that the concept and method of HRI, the cornerstone of the HPM decision-making system, generated interest in the study of other contexts and with different threats in Latin America, which is an indication of the usefulness of this index in risk management in complex systems. Among the most significant challenges that lie ahead for HPM to contribute to transforming a society based on the use of pesticides by a more equal society, more environment friendly and economically viable, are: produce agricultural products and conserve natural resources at the same time; achieve profitability irrespective of agrochemicals; and empower people to improve their situation. Many voices cry out because it is so.

References

Altieri, M. A., & Nicholls, C. I. (2005). *Agroecology and the search for a truly sustainable agriculture* (1st ed.). Mexico: United Nations Environment Programme, Environmental Training Network for Latin America and the Caribbean.

Álvarez Morales, Y. (2015). *Evaluación de indicadores de sustentabilidad agroecológica en sistemas de producción agrícola de Baja California Sur, México*. Tesis de doctorado, Centro de Investigaciones Biológicas del Noroeste, S.C. La Paz, Baja California Sur, Mexico.

Anderson, F. (2018). ¡Soberanía alimentaria ya! Coordinación Europea Vía Campesina. Belgium.

Andrews, K. L. (1989). Introducción a los conceptos del manejo integrado de plagas. In K. L. Andrews & J. R. Quezada (Eds.), *Manejo integrado de plagas insectiles en la agricultura: Estado actual y futuro* (pp. 3–20). El Zamorano, Honduras: Escuela Agrícola Panamericana.

Andrews, K. L. (2006). Prólogo. In A. Castillo, J. Gómez, H. Morales, J. Toledo, & R. Jarquín, (Eds.). Resúmenes *del X Congreso Internacional de Manejo Integrado de Plagas y Agroecología* (27 al 29 de septiembre de 2006). Tapachula, Chiapas, Mexico.

Aselage, J., & Johnson, D. T. (2009). From IPM to organic and sustainable agriculture. In E. B. Radcliffe, W. D. Hutchinson, & R. E. Cancelado (Eds.), *Integrated Pest Management: Concepts, tactics, strategies and case studies* (pp. 489–505). Cambridge: Cambridge University Press.

Barrera, J. F. (2006). Manejo holístico de plagas: Hacia un nuevo paradigma de la protección fitosanitaria. In J. Pohlan, L. Soto, & J. Barrera (Eds.), *El cafetal del futuro: Realidades y Visiones* (pp. 63–82). Aachen: Shaker Verlag.

Barrera, J. F. (2009). La necesidad del enfoque holístico en el manejo de plagas. In J. C. Arrivillaga, M. El Souki, & B. Herrera (Eds.), *Enfoques y temáticas en entomología. XXI Congreso Venezolano de Entomología* (pp. 1–28). Caracas: Sociedad Venezolana de Entomología.

Barrera, J. F. (2020). Holistic Pest Management. In D. Perez-Staples, F. Diaz-Fleischer, P. Montoya, & M. T. Vera (Eds.), *Area-wide management of fruit fly pests* (pp. 275–293). Boca Raton: CRC Press, Taylor & Francis Group.

Barrera, J. F., Herrera, J., & Gómez, J. (2007). Riesgo-vulnerabilidad hacia la broca del café bajo un enfoque de manejo holístico. In J. F. Barrera, A. García, V. Domínguez, & C. Luna (Eds.), *La Broca del Café en América Tropical: Hallazgos y Enfoques* (pp. 131–141). Mexico: Sociedad Mexicana de Entomología y El Colegio de la Frontera Sur.

Barrera, J. F., Gamboa, W., Gómez, J., et al. (2011). Método holístico para la toma de decisiones en el manejo de plagas. In *Simposio Estado del arte del Manejo Ecológico de Plagas en América Latina*. Tercer Congreso Latinoamericano de Agroecología, August 17–19, Oaxtepec, Mexico.

Barrera, J. F., Valle, J., Gómez, J., et al. (2017). *Manejo holístico de plagas en zonas cafetaleras: Concepto y método*. Mexico: El Colegio de la Frontera Sur.

Benbrook, C. M., Groth, E., III, Halloran, J. M., et al. (1996). *Pest management at the crossroads*. Yonkers: Consumers Union of United States, Inc..

Bentley, J., & Andrews, K. (1996). Through the roadblocks: IPM and Central American smallholders. Sustainable Agriculture and Rural Livelihoods Programme. International Institute for Environment and Development. Gatekeeper Series No. 56. London.

Bottrell, D. G. (1979). *Integrated pest management: Council of environmental quality*. Washington, DC: Government Printing Office.

Braun, A. R., Thiele, G., & Fernandez, M. (2000). *Farmer Field Schools and Local Agricultural Research Committees: Complementary platforms for integrated decision-making in sustainable agriculture*. Agricultural Research and Extension Network, ODI. AgREN Network Paper No. 105. London.

Capra, F. (1998). *La trama de la vida, una nueva perspectiva de los sistemas vivos*. Barcelona: Ed. Anagrama SA.

Carpio Sacoto, J. E., & Carpio Sacoto, L. M. (2014). *Determinación de estrategias sociales y ecológicas de adaptación al cambio climático implementadas por los agricultores de las cuatro zonas agroecológicas de la Parroquia San Joaquín*. Tesis de maestría, Universidad Politécnica Salesiana, Cuenca, Equator.

Castro-Triana, A. M. (2018). *Estudio sobre la repelencia y la atracción en la broca del café como herramienta para el manejo agroecológico en los cafetales colombianos*. Tesis de doctorado, Universidad de Antioquia, Medellín, Colombia.

Drucker, P. F. (1993). *La sociedad postcapitalista*. Spain: Ed. Apóstrofe.

Enkerlin, W. R., Gutiérrez Ruelas, J. M., Pantaleon, R., et al. (2017). The Moscamed Regional Programme: Review of a success story of area-wide sterile insect technique application. *Entomologia Experimentalis et Applicata, 164*, 188–203.

Gazzano, I., & Achkar, M. (2015). *Amenaza, Vulnerabilidad y Riesgo: Estrategias de respuesta de ganaderos familiares en el área protegida Esteros de Farrapos-Uruguay*. In Memorias del V Congreso Latinoamericano de Agroecología, La Plata, Argentina.

Gazzano, I., & Achkar, M. (2016). Conflictos de las transformaciones territoriales: Ganaderos frente a la intensificación agraria en Esteros de Farrapos Uruguay. *Revista Iberoamericana de Economía Ecológica, 26*, 109–121.

Gazzano Santos, M. I. (2014). *Viabilidad de la ganadería familiar en áreas protegidas de humedales, en un contexto sinérgico de intensificación agraria e inundaciones: Parque Nacional Esteros de Farrapos-Uruguay*. Doctoral thesis, Universidad de Córdoba, Spain.

Gazzano, I., Altieri, M., Achkar, M., et al. (2015). Holistic Risk Index: A case study of cattle producers in the protected área of Farropos Estuaries-Uruguay. *Agroecology and Sustainable Food Systems, 39*(2), 209–223.

Gazzano, I., Altieri, M., Achkar Borrás, M., et al. (2016). Riesgo y resiliencia de productores ganaderos familiares en el área protegida Parque Nacional Esteros de Farrapos, Uruguay. *Agrociencia Uruguay, 20*(1), 51–60.

Gray, M. E., Ratcliffe, S. T., & Rice, M. E. (2009). The IPM paradigm: Concepts, strategies and tactics. In E. B. Radcliffe, W. D. Hutchinson, & R. E. Cancelado (Eds.), *Integrated Pest Management: Concepts, tactics, strategies and case studies* (pp. 1–13). Cambridge: Cambridge University Press.

Griffin, R. (2012). Basic concepts in risk analysis. In C. Devorshak (Ed.), *Plant pest risk analysis: Concepts and application* (pp. 7–18). Wallingford: CAB International.

Henao-Salazar, A. (2013). Propuesta metodológica de medición de la resiliencia agroecológica en sistemas socio-ecológicos: Un estudio de caso en los Andes colombianos. *Agroecología, 8*(1), 85–91.

Hendrichs, J., Kenmore, P., Robinson, A. S., et al. (2007). Area-wide integrated pest management (AW-IPM): Principles, practice and prospects. In V. MJB, A. S. Robinson, & J. Hendrichs (Eds.), *Area-wide control of insect pests* (pp. 3–33). Dordrecht: Springer.

Herrera, O. B., Parra, M., Livscovsky, I., et al. (2019). Lifeways and territorial innovation: Values and practices for promoting collective appropriation of territory. *Community Development Journal, 54*(3), 427–445.

Higley, L. G., & Pedigo, L. P. (1996). The EIL concept. In L. G. Higley & L. P. Pedigo (Eds.), *Economic thresholds for Integrated Pest Management* (pp. 9–21). Lincoln: University of Nebraska Press.

Higley, L. G., & Peterson, R. K. D. (2009). Economic decision rules for IPM. In E. B. Radcliffe, W. D. Hutchinson, & R. E. Cancelado (Eds.), *Integrated Pest Management: Concepts, tactics, strategies and case studies* (pp. 25–32). Cambridge: Cambridge University Press.

Hokkanen, H. M. T. (2015). Integrated pest management at the crossroads: Science, politics, or business (as usual)? *Arthropod-Plant Interactions, 9*, 543–545.

Klassen, W. (2000). Area-wide approaches to insect pest management: History and lessons. In K. H. Tan (Ed.), *Proceedings: Area-wide control of fruit flies and other insect pests. International conference on area-wide control of insect pests, and the 5th international symposium on fruit flies of economic importance, May 28–June 5, 1998, Penang, Malaysia* (pp. 21–38). Pulau Pinang, Malaysia: Penerbit Universiti Sains Malaysia.

Kogan, M. (1998). Integrated pest management: Historical perspectives and contemporary developments. *Annual Review of Entomology, 43*, 243–270.

Levins, R. (2007). From simple IPM to the management of agroecosystems. In M. Kogan & P. Jepson (Eds.), *Perspectives in ecological theory and Integrated Pest Management* (pp. 45–64). Cambridge: Cambridge University Press.

Lewis, W. J., van Lenteren, J. C., Phatak, S. C., et al. (1997). A total system approach to sustainable pest management. *Proceedings of the National Academy of Sciences of the United States of America, 94*, 12243–12248.

Liebig, T., Jassogne, L., Rahn, E., et al. (2016). Towards a collaborative research: A case study on linking science to farmers' perceptions and knowledge on Arabica coffee pests and diseases and its management. *PLoS One, 11*(8), e0159392.

Lindquist, D. A. (2000). Pest management strategies: Area-wide and conventional. In K. H. Tan (Ed.), *Proceedings: Area-wide control of fruit flies and other insect pests. In International conference on area-wide control of insect pests, and the 5th international symposium on fruit flies of economic importance, May 28–June 5, 1998, Penang, Malaysia* (pp. 13–19). Pulau Pinang, Malaysia: Penerbit Universiti Sains Malaysia.

Machado-Vargas, M. M. (2017). Marco conceptual para evaluar los niveles de resiliencia socio-ecológica, aplicación en estudios de caso en pequeños productores de café. In C. I. Nicholls & M. A. Altieri (Eds.), *Nuevos caminos para reforzar la resiliencia agroecológica al cambio climático* (pp. 46–53). Berkeley: Sociedad Científica Latinoamericana de Agroecología (SOCLA) y Red Iberoamericana de Agroecología para el Desarrollo de Sistemas Agrícolas Resilientes al Cambio Climático (REDAGRES).

Machado-Vargas, M. M., Nicholls-Estrada, C. I., & Ríos-Osorio, L. A. (2018). Social-ecological resilience of small-scale coffee production in the Porce river basin, Antioquia (Colombia). *Idesia, 36*(3), 141–151.

Menalled, F. D., Landis, D. A., & Dyer, L. E. (2004). Research and extension supporting ecologically based IPM systems. *Journal of Crop Improvement, 11*(1–2), 153–174.

Meneses-Ospina, E. (2015). Bases para el manejo agroecológico de *Eurhizococcus colombianus* Jakubski (Hemiptera: Margarodidae) en cultivos de mora del Oriente Antioqueño. Tesis de doctorado, Universidad Nacional de Colombia, Facultad de Ciencias Agrarias, Medellín, Colombia.

Montalba, R., García, M., Altieri, M., et al. (2013). Utilización del Índice Holístico de Riesgo (IHR) como medida de resiliencia socioecológica a condiciones de escasez de recursos hídricos. Aplicación en comunidades campesinas e indígenas de la Araucanía, Chile. *Agroecología, 8*(1), 63–70.

Montalba, R., Fonseca, F., García, M., et al. (2015). Determinación de los niveles de riesgo socioecológico ante sequías en sistemas agrícolas campesinos de La Araucanía chilena. Influencia de la diversidad cultural y la agrobiodiversidad. *Papers Revista de Sociología, 100*(4), 607–624.

Morin, E. (2000). *La mente bien ordenada: Pensar la reforma, reformar el pensamiento* (185 p). Barcelona: Ed. Siex Barral S.A.

Morse, S. (2009). IPM: Ideals and realities in developing countries. In E. B. Radcliffe, W. D. Hutchinson, & R. E. Cancelado (Eds.), *Integrated Pest Management: Concepts, tactics, strategies and case studies* (pp. 458–470). Cambridge: Cambridge University Press.

Morse, S., & Buhler, W. (1997). *Integrated pest management: Ideals and realities in developing countries*. London: Lynne Rienner Publishers.

Munyuli, T., Cihire, K., Rubabura, D., et al. (2017). Farmers' perceptions, believes, knowledge and management practices of potato pests in South-Kivu Province, eastern of Democratic Republic of Congo. *Open Agriculture, 2*, 362–385.

Nicholls, C. I., Altieri, M. A., & Vazquez, L. (2016). Agroecology: Principles for the conversion and redesign of farming systems. *Journal of Ecosystem & Ecography, S5*, 010.

Norris, R. F., Caswell-Chen, E. P., & Kogan, M. (2003). *Concepts in Integrated Pest Management*. Upper Saddle River: Prentice Hall.

NRC (National Research Council). (1996). *Ecologically based pest management: New solutions for a new century*. Washington, DC: National Academy Press.

Parsa, S., Morse, S., Bonifacio, A., et al. (2014). Obstacles to integrated pest management adoption in developing countries. *Proceedings of the National Academy of Sciences of the United States of America, 111*(10), 3889–3894.

Prokopy, R. J., & Kogan, M. (2003). Integrated Pest Management. In V. H. Resh & R. T. Cardé (Eds.), *Encyclopedia of insects* (pp. 4–9). New York: Academic.

Reyes Robles, C., Barrera, J. F., Moreno Reséndez, A., et al. (2017). Índice holÍstico de riesgo para la toma de decisiones en el manejo de la roya *Hemileia vastatrix* en cafetales de Honduras. In *XXIII Simposio Latinoamericano de Caficultura*. El Programa Cooperativo Regional para el Desarrollo Tecnológico y Modernización de la Caficultura (PROMECAFE) e Instituto Hondureño del Café (IHCAFE), August 23–26, 2017, San Pedro Sula, Honduras

Royer, T. A., Mulder, P. G., & Cuperus, G. W. (1999). Renaming (redefining) Integrated Pest Management: Fumble, pass, or play? *American Entomologist, 45*(3), 136–139.

Savory, A., & Butterfield, J. (1999). *Holistic management, a new framework for decision making* (2nd ed., 616 p). Washington, DC: Island Press.

Scoones, I. (2017). *Medios de vida sostenibles y desarrollo rural*. Barcelona: Icaria Editorial, S.A..

Segura, H. R., Barrera, J. F., Nazar, A., et al. (2004). Farmers' perceptions, knowledge and management of the coffee pests and diseases and their natural enemies in Chiapas, Mexico. *Journal of Economic Entomology, 97*, 1491–1499.

Stenberg, J. A. (2017). A conceptual framework for Integrated Pest Management. *Trends in Plant Science, 22*(9), 759–769.

Stern, V. M., Smith, R. F., van den Bosch, R., et al. (1959). The integrated control concept. *Hilgardia, 29*(2), 81–101.

Thrupp, L. A. (1996). Overview. In L. A. Thrupp (Ed.), *New partnerships for sustainable agriculture* (pp. 1–38). Washington, DC: World Resources Institute.

van de Fliert, E., & Braun, A. R. (2002). Conceptualizing integrative, farmer participatory research for sustainable agriculture: From opportunities to impact. *Agriculture and Human Values, 19*, 25–38.

Waage, J. K. (1992). Biological control in the year 2000. In A. A. S. A. Kadir & H. S. Barlow (Eds.), *Pest management and the environment in 2000* (pp. 329–340). Wallingford: CAB International.

Walters, M. L., Sequeira, R., Staten, R., et al. (2009). Eradication: Strategies and tactics. In E. B. Radcliffe, W. D. Hutchinson, & R. E. Cancelado (Eds.), *Integrated Pest Management: Concepts, tactics, strategies and case studies* (pp. 298–308). Cambridge: Cambridge University Press.

Index

J. F. Barrera, *Beyond IPM: Introduction to the Theory of Holistic Pest Management*, Sustainability in Plant and Crop Protection 15,
https://doi.org/10.1007/978-3-030-43370-3

I

K

L

Q

R

MIX
Papier aus verantwortungsvollen Quellen
Paper from responsible sources
FSC® C105338

If you have any concerns about our products,
you can contact us on
ProductSafety@springernature.com

In case Publisher is established outside the EU,
the EU authorized representative is:
Springer Nature Customer Service Center GmbH
Europaplatz 3, 69115 Heidelberg, Germany

Printed by Libri Plureos GmbH
in Hamburg, Germany